WITHDRAWAL

Citizenship
in an Age
of Science

Pergamon Titles of Related Interest

De Volpi/Marsh/Postol/Stanford   The H-Bomb, *The Progressive,* And Government Secrecy
Golden   Science Advice To The President
Gordon/Gerjuoy/Anderson   Life-Extending Technologies
Starr/Ritterbush   Science, Technology and the Human Prospect

Related Journals*

Fundamenta Scientiae
Futurics
Technology In Society

*Free specimen copies available upon request.

PERGAMON POLICY STUDIES On SCIENCE AND TECHNOLOGY

Citizenship in an Age of Science
Changing Attitudes Among Young Adults

Jon D. Miller
Robert W. Suchner
Alan M. Voelker

Pergamon Press
NEW YORK • OXFORD • TORONTO • SYDNEY • FRANKFURT • PARIS

Pergamon Press Offices:

U.S.A.   Pergamon Press Inc., Maxwell House, Fairview Park,
                   Elmsford, New York 10523, U.S.A.

U.K.   Pergamon Press Ltd., Headington Hill Hall,
                   Oxford OX3 0BW, England

CANADA   Pergamon of Canada, Ltd., Suite 104, 150 Consumers Road,
                   Willowdale, Ontario M2J 1P9, Canada

AUSTRALIA   Pergamon Press (Aust.) Pty. Ltd., P.O. Box 544,
                   Potts Point, NSW 2011, Australia

FRANCE   Pergamon Press SARL, 24 rue des Ecoles,
                   75240 Paris, Cedex 05, France

FEDERAL REPUBLIC   Pergamon Press GmbH, Hammerweg 6, Postfach 1305,
OF GERMANY   6242 Kronberg/Taunus, Federal Republic of Germany

Copyright © 1980 Pergamon Press Inc.

Library of Congress Cataloging in Publication Data

Miller, Jon D   1941-
    Citizenship in an age of science.

    (Pergamon policy studies)
    Bibliography: p.
    Includes index.
    1. Science—Social aspects—United States. 2. Sci-
ence and state—United States. I. Suchner, Robert W.,
1944-   joint author. II. Voelker, Alan M., joint
author. III. Title.
Q175.52.U5M54 1980   306'.4   79-27861
ISBN 0-08-024662-1

Printed in the United States of America

To

Gabriel A. Almond

whose basic theoretical insights
provide the foundation for this work

Contents

Foreword   xi

Acknowledgements   xv

      Section I: THE BASIC CONCEPTS

Chapter

  1   THE EVOLUTION OF ORGANIZED SCIENCE   3
        The Practice of Science in Organized
          Settings   4
        The Public Support of Organized Science   9
        The Public Recognition of Organized
          Science   12
        The Outlook for the Future   16

  2   THE PUBLIC UNDERSTANDING OF
          ORGANIZED SCIENCE   19
        Opinions, Attitudes, and Ideologies   21
        General Political Attitudes   22
        The Adult Understanding of Organized
          Science   24
        The Preadult Understanding of
          Organized Science   27
        A Question of Saliency   29

  3   THE DEVELOPMENT OF ISSUE ATTENTIVENESS   33
        The Development of General
          Political Interest   33
        The Development of Issue Attentiveness   43
        Attentiveness to Organized Science   48

Section II: THE DATA BASE

Chapter

    4   THE 1978 NATIONAL PUBLIC AFFAIRS STUDY   55
         The NPAS Questionnaire   55
         Population and Sample   57
         Data Collection   61
         Weighting the Sample   62
         Coding Procedure and Replacement of
            Missing Data   64
         Index Construction   65

    Section III: ATTENTIVENESS TO ORGANIZED SCIENCE

Chapter

    5   THE STRUCTURE AND DEVELOPMENT OF
              INTEREST IN ORGANIZED SCIENCE   73
         The Concept of Issue Interest   73
         Developmental Patterns   84
         Specialized Interests   88

    6   THE LEVELS AND SOURCES OF SCIENCE
              AND TECHNOLOGY INFORMATION   93
         Current Information About Science
            and Technology Issues   93
         Information Acquisition   99

    7   THE DEVELOPMENT OF ATTENTIVENESS   119
         The Attentiveness Typology   119
         The Structure of Attentiveness   122
         The Development of Attentiveness   128

    8   THE IMPACT OF ATTENTIVENESS ON SCIENCE
              AND TECHNOLOGY POLICY VIEWS   135
         General Attitudes Toward Science
            and Technology   136
         Specific Science and Technology
            Policy Views   155

    Section IV: THE SOURCES OF ATTENTIVENESS

    9   THE FAMILY   167
         Socioeconomic Status   168
         Educational Aspirations   170
         Occupational Aspirations   174
         Sex-Role Socialization   180
         Religious Beliefs and Participation   183

Family Politicization   190
A Process Model of Family Effects   193

10   SCHOOL AND PEERS   201
     The Base Family Effects Model   202
     Academic Achievement and Science
        Coursework   203
     Politicization   208
     Life Goals and Occupational Preference 212

11   PERSONALITY   219
     Personality Traits Associated With
        Attentiveness to Science   219
     Interest, Knowledge, and Exposure
        to Science Information   241
     A Model of Personality Factors
        Associated With Attentiveness   246

12   ATTENTIVENESS AND GENERAL POLITICAL
        INTEREST   263
     A Measure of General Political
        Interest   263
     A Comparison of Three Models   266

13   A DEVELOPMENTAL MODEL   275
     The Independent Variables   275
     A Logit Model   278
     A Path Model   280

Section V: THE IMPLICATIONS OF ATTENTIVENESS

14   CITIZENSHIP IN AN AGE OF SCIENCE   289
     The Role of Attentive Publics   289
     The Attentive Public for Organized
        Science   300

Appendix

A   The 1978 NPAS Instruments   303

B   Log-linear Analysis--Logic and
        Procedures   321

Notes   335

References   339

Index   357

About the Authors   361

Foreword

Most important issues facing the world today are scientific and technical in character. This is true across a wide spectrum of needs and dangers confronting the human race, from war through economic development to the generation of new knowledge. Consequently, we find that the average citizen, if not directly, then indirectly through elected representatives, must pass judgments on public policy issues of increasing complexity.

Citizenship in an Age of Science addresses the equipment that young Americans today have at their disposal in coping with these complexities. The authors' words bring to mind a team of medical specialists evaluating the condition of a patient before cures can be prescribed. As diagnosticians, Miller, Suchner, and Voelker masterfully analyze a complex set of symptoms affecting the American political and educational psyche.

For those readers who have more than a passing interest in science and technology policy or the public understanding of science, the authors confirm a lot of things that might have been suspected all along. It's not too surprising to find out that most of those attentive to science are also attentive to technology. Nor do we raise eyebrows over confirmation of the importance of an individual's family background, gender, occupational and educational aspirations, or self-esteem in predicting who will be attentive to science and who will not.

But conventional wisdom is not always confirmed. There are surprises, such as the irrelevance of general academic achievement to interest in organized

science and the relatively benign personality traits
associated with those attentive to science. Disap-
pointments crop up, too, such as indications that
science curriculum development and course improvement in our schools and universities have been less
effective in contributing to attentiveness to science than we might wish.

This is an extremely valuable work for a number
of reasons. The precise and carefully crafted definitions provide a common vocabulary for discussion
among social scientists and other experts. Quanti-
fication of effects that were suspected but not
proven in the past removes nagging doubts and establishes the basis for evaluations of past programs
and more rational allocation of future resources
dedicated to sensitizing our young people to science
and technology.

Perhaps the greatest value of this work lies in
its lessons for the future. There are several. The
authors point out the need for more information,
particularly longitudinal data. The rapid changes
in attitudes toward economic and political issues
exhibited by college students in the last few years
demonstrate how short-lived are presumably well-
established norms. Based on the information now
available, one sees a great need for members of the
scientific and engineering communities to exert
great efforts to communicate with our nonattentive
brethren, including decision-makers in the political and corporate spheres. The finding that a substantial portion of our young adults who do not meet
the criteria for attentiveness to science but do
have a demonstrable interest serves to define a
large population of potential attentives. By char-
acterizing these individuals, the authors provide a
road map for locating them.

But the question arises immediately, "why both-
er?" The answers are several, and affect society as
a whole as well as the scientific and engineering
communities. The superb discussion of risk and benefit attitudes as a function of attentiveness to
science takes on added importance as we project a
future with technologies that are laden with insepa-
rable risks and benefits. The actuarial approach to
decision-making is substantially more prevalent
among those attentive to science, and this characteristic clearly is vital to the political and social stability necessary for long-term prosperity
and even for survival of our way of life. It is
also encouraging to find that those enamored of

science, often thought of as pointy-headed and im-
practical, are imbued with at least as much common
sense as are historians or auto mechanics.
     The tangible benefits of investments in re-
search and development are better recognized by
those attentive to science than those nonattentive.
This has implications for the national economy that
lie beyond immediate cash-flow considerations of
scientists and engineers in their R&D projects. As
William Nordhaus eloquently pointed out while a
member of the Council of Economic Advisors in 1977,
the social rate of return on R&D investments is ex-
ceedingly high but more uncertain than other private
investments. Diversification coupled with an actu-
rial viewpoint eliminates the uncertainty for society as a whole. Again we see that it is the person
attentive to science who has the mind-set needed to
grapple with so complex but so practical an issue.
     Citizenship in an Age of Science provides an
authoritative basis for future studies. But it does
a lot more. It implies an unfinished agenda for the
scientist and engineer. I would expect renewed interest and expanded activities directed toward increasing the proportion of Americans attentive to
science and technology by elements of organized science such as the disciplinary scientific and engineering societies. These institutional initiatives
are important, but even more so is a grassroots recognition by scientists and engineers of the need for
better communication with the public and the con-
comitant development of a better habit of public
service. The result will surely be a public more
attentive to science and technology and a society
better equipped to make the complex decisions facing
our nation and the world in the years to come.

                         J. Thomas Ratchford
                         American Association for
                            the Advancement of
                            Science
                         February 1980

Acknowledgments

The design, execution, and analysis of a national attitude study is a humbling experience. The process of acknowledging the assistance received demonstrates the dependence of the principal investigators on the skills, talents, and patience of numerous others.

Our first and most basic debt is to Professor Gabriel Almond, whose early work in the structure of public attitudes toward foreign policy provided the intellectual stimulus and foundation for this work. In recognition of this debt, we have dedicated this monograph to him.

Our second major debt is to the National Science Foundation for the financial support of a major portion of the cost of this study through grant number SED77-18491. In addition to the funds provided, the advice and assistance of Ray Hannapel, Andrew Molnar, Mary Budd Rowe, and Joseph Lipson were invaluable. All opinions, findings, conclusions, or recommendations expressed in this publication are those of the authors and do not necessarily express the views of the National Science Foundation or its staff.

Third, we wish to acknowledge the assistance of the approximately 4700 young persons who each contributed an hour of their time to complete our instrument and the principals, curriculum coordinators, teachers, college registars, and deans who made the sample selection process possible. We would have preferred to name many of the individuals whose help in sample selection went far beyond the call of duty, but our commitment to the confidentiality of the participating school systems prohibits a full expression of our gratitude.

Throughout the long months of data collection, analysis, writing, and manuscript preparation, we have been served by a most energic and capable collection of graduate and research assistants. The data collection and editing processes were aided by the work of Ray Chang, Robert Hennig, Gay McMorrow, Mac Robinson, and innumerable undergraduates from Northern Illinois University and other young people from the DeKalb County, Illinois, CETA program. The data analysis segment of the work was aided by the work of John Kloyp, Joel Kallich, Georgette Rocheleau, Jan McConeghy, Joan Kalmanek, and Beth Nagle. The final manuscript would not have been completed without the word processing skills of Jan McConeghy, Janis Gorski, and Tom Barrington and the art work of Hilkka Itkonen.

The patience, understanding, and assistance of Linda Schwarz, Kay Van Mol, Debbie Silvestri, and other members of the staff of the Graduate School Office of Research, in whose facilities much of the work was performed, made the tasks much more tolerable. And without the coordinative assistance of Caroline Wood, Dean Miller's secretary, the timely completion of this work would have been problematic at best.

Inevitably, the conduct of a study of the proportions of the 1978 National Public Affairs Study creates conflicts in time commitments and facility availability for the principal investigators and their colleagues. We wish to acknowledge the kind support of Dr. James A. Rutledge, Dean of the Graduate School; Dr. Eleanor Godfrey, former Chair of the Department of Sociology; and Dr. Charles Sloan, former Chair of the Department of Elementary Education. Without their support and understanding, the work reported herein would not have been possible.

Last, but certainly not least, we wish to thank our families for their patience and understanding for the long hours of evening and weekend time devoted to this project.

                          Jon D. Miller
                          Robert S. Suchner
                          Alan M. Voelker

                          DeKalb, Illinois
                          February, 1980

Citizenship
in an Age
of Science

I
The Basic Concepts

1 The Evolution of Organized Science

As a preface to understanding the development of public attitudes toward science, it is necessary to focus first on the nature of the institutions and practices of science, both of which have undergone a substantial evolution in the twentieth century. Too often, public opinion analysts have focused on changes in the public's attitudes toward science, implicitly assuming that science was relatively unchanging. In fact, it is likely that the institutions and practices of science in the United States have changed more substantially during the last century than have the public attitudes toward it.

While the evolution of modern science is a vast subject domain, three aspects of change emerge as critical to an understanding of public attitudes. First, the practice of science has increasingly depended upon complex combinations of specialized personnel and facilities, and this has meant that most modern science is practiced in organizations. Further, the nature of the resource requirements in many disciplines demands relatively large organizations. The growth of the institutional basis of science is perhaps no greater proportionately than similar changes in manufacturing, retail distribution, or transportation, but it is important to understand the nature, magnitude, and implications of this change. The term "organized science" (1) is a useful reminder of this change.

Second, the growth of scientific institutions has demanded increasingly higher levels of public support. While a substantial amount of applied research is supported by for-profit corporations, the

speculative nature of basic research has kept it
largely in the public and not-for-profit realm.
Even those economists who would infuse free enterprise into education and postal services have
stopped short of arguing for the support of basic
scientific research by free enterprise alone. The
support of scientific research is now a major object
of federal expenditures. In the competition for
federal dollars, science must compete with other social needs and is increasingly judged in a comparative context.

Third, as the impact and cost of organized science have grown, it has increasingly become the subject of public scrutiny and, in some cases, public
regulation. This is hardly surprising. There is
little reason for the public to devote its attention
to groups that make insignificant demands on the
public purse and whose activities have no direct
public consequences, especially negative consequences. Where science may have met such criteria
at the turn of the century, it has clearly become an
appropriate object of public awareness in the last
three decades. The combination of pervasive scientific impact on American society and the greater
public awareness of organized science has led to an
increasing number of proposals to assert a public
regulatory power over scientific and technological
activities, especially the latter. Accustomed to
little external regulation, many leaders in the scientific community have resisted even minimal public
scrutiny, and a debate over the proper role of public and peer regulation is well underway in the
inner circles of American science and politics.

In view of the import of these three evolution-
ary patterns of organized science, it is appropriate
to examine each of these trends in greater detail as
a preface to our analysis of public attitudes toward
organized science.

THE PRACTICE OF SCIENCE IN ORGANIZED SETTINGS

The single most important change in the practice of
science in the twentieth century has been the emergence of "big science," as characterized by team research and large and expensive physical facilities.
Individual creativity and intellect continue to be
the most important ingredients in organized science,
but an ever larger portion of scientific and techno-

logical research requires complex combinations of personnel and physical resources to exercise the basic
intelligence and creativity that drives the processes of discovery and innovation.
        Weinberg (1972) observes that:

        Nineteenth-century science was mainly con-
        ducted by geographically isolated, though
        intellectually interacting individuals; much
        of today's science is conducted by large in-
        terdisciplinary teams. These teams often
        center around pieces of expensive equipment
        and are then said to be part of "big sci-
        ence." Team science is characteristically
        conducted in the large multipurpose scien-
        tific laboratory, an institution that is
        predominately a phenomenon of World War II
        and after....

        The emergence of the large interdisciplinary
        scientific team as the landmark of science
        can be traced to at least three separate de-
        velopments. First is the extraordinary
        growth of science and the resulting increase
        in the amount of scientific information pro-
        duced; second is the emergence and institu-
        tionalization of applied science; third, and
        possibly most important, is the increasing
        complexity of scientific machinery (p. 114).

Variously described, the same process has been discussed by Reagan (1969), Struve (1971), Horniq
(1971), Brooks (1968), and Pierce and Tressler
(1964). Taylor (1973) has described a very similar
process of institutionalization in the United King-
dom. In view of the importance of this process to
an understanding of organized science, it is useful
to examine briefly each of the three factors noted
by Weinberg.
        The scientific information explosion of the
three decades following the end of the Second World
War has been widely noted and studied. Menard
(1971) has demonstrated that the growth of scientific publication has been exponential during the period since 1945 in a number of scientific disciplines.
King (1975) reports that it took Chemical Abstracts
30 years to publish its first million abstracts, 18
years to publish its second million, eight years to
publish its third million, four years to publish its

fourth million, and less than three years to publish
its fifth million.

With the growth of information, specialization
has been a necessary result. As the segment of a
discipline that one person can master grows narrower
and narrower, the growth of research teams becomes
mandatory if the necessary range of expertise is to
be brought to any given research problem. This
basic process of specialization is ultimately reflected in the increasingly specialized graduate
education provided to new entrants into the field.
No longer does an individual earn a doctorate in
chemistry and expect to be minimally knowledgeable
in all of the facets of the discipline, but rather
the graduate student in chemistry becomes an expert
in a segment of the broader discipline. Wiener
(1948) observed at the beginning of the postwar
information explosion that the level of specialization was already substantial, and that:

> Since Leibniz, there has perhaps been no man
> who has had a full command of all of the in-
> tellectual activity of his day. Since that
> time, science has been increasingly the task
> of specialists, in fields which show a ten-
> dency to grow progressively narrower....
> Today, there are few scholars who can call
> themselves mathematicians or physicists or
> biologists without restriction. A man may
> be a topologist or an acoustician or a cole-
> opterist. He will be filled with the jargon
> of his field and will know all its litera-
> ture, but, more frequently than not, he will
> regard the next subject as something belong-
> ing to his colleague three doors down the
> corridor (p. 8).

In a parallel manner, applied science has grown
to substantial dimensions and has been institutionalized at least as much -- and probably more -- than
basic science research. In their history of science
in New Jersey, Pierce and Tressler (1964) provide an
unusual and important portrait of the growth and institutionalization of applied science. The contrast
between the small, early workshops of Thomas Edison
in Newark and Menlo Park and the current massive
Bell Laboratories complex at Murray Hill illustrates
the distance covered in the last century.

The extensive involvement of the scientific and
technological communities in the Second World War

marked a turning point in the magnitude of applied-
science projects and direct federal involvement in
the process (Stewart, 1948). President Roosevelt's
creation of an Office of Scientific Research and Development (OSRD) in his own executive office provided a leadership role for the scientific community,
and the ultimate success of the Manhattan Project
generated substantial public and governmental expectations for the utility of science in the postwar
years. Vannevar Bush, the head of OSRD, led a national movement to continue the wartime national
laboratories (like those in Los Alamos and Oak
Ridge) and to utilize the discoveries of basic science for a broad array of societal needs. Stressing
the economic advantages, Bush (1945) argued:

> One of our hopes is that after the war there
> will be full employment. To reach that
> goal, the full creative and productive ener-
> gies of the American People must be releas-
> ed. To create more jobs we must make new
> and better and cheaper products. We want
> plenty of new, vigorous enterprises. But
> new products and processes are not born
> full-grown. They are founded on new princi-
> ples and new conceptions which in turn re-
> sult from basic scientific research. Basic
> scientific research is scientific capital.
> Clearly, more and better scientific research
> is one essential to the achievement of our
> goal of full employment (p. 2).

The launching of the first Sputnik satellite by
the Soviet Union in 1957 marked the beginning of a
decade of massive federal support for the application of science and technology to space exploration.
Ramo (1970) observes that the Apollo Program was the
largest single project in the nation's history,
larger than "the atomic bomb project, the intercon-
tinental ballistic missile project, the Panama
Canal, the Hoover Dam, and, so far at least, the
federal government's war on poverty" (p. 43).
It is reasonable to project that the present
federal concern for energy resources will spur another set of extensive applied-science projects.
In fact, several of the national laboratories that
emerged from the Second World War era have turned
their attentions to energy problems. Weinberg
(1967) has argued that large-scale applied-science
projects in nuclear energy and similar areas are

inappropriate for the modern university structure.
He has contended that missions of that type are
ideal for national laboratories like Oak Ridge.
More recently, President Carter has compared the new
emphasis on alternative energy sources to the space
program.

One of the major justifications for the institutionalization of both basic science and applied
research has been the large and expensive nature of
the instrumentation needed. Weinberg (1972) com-
pares the modern scientific research team to the
early explorers of a previous century:

> To a degree, we would have to regard the
> great explorers as geographers, and hence
> scientists of sorts. Their enterprises were
> on a grand scale by the standards of their
> time; they required large teams and much
> money. And at least Columbus among them
> politicked with John of Portugal and Queen
> Isabella in much the same way that a promot-
> er of a large accelerator must now politick
> with the Atomic Energy Commission or the
> National Science Foundation, or even with
> the President himself, to sell his project.
>
> Many of today's explorations in basic sci-
> ence involve such elaborate and expensive
> pieces of hardware that the whole enter-
> prise requires the same mobilization of
> resources as was required by the explorers
> (p. 120).

The dependence on large and expensive equipment
is most advanced in physics and astronomy, but similar patterns are dominant in many university and industrial laboratories. The demand of researchers in
numerous scientific disciplines for electron micro-
scopes, nuclear magnetic resonators (NMRs), and similar expensive items has significantly reduced the
number of institutions at which first-rate research
can be conducted. Struve (1971) has described the
reduction in the number of centers for astronomy
research as a result of the need for larger and more
advanced telescopes.

Similar patterns have emerged in applied scientific research as well as the more basic work. The
space program is perhaps the best example. The cost
of launching a space satellite or a larger vehicle
is prohibitive for any group other than the federal

government and thus the National Aeronautics and
Space Administration (NASA) has played a broker role
in seeking to utilize the expertise of numerous university and industrial laboratories while maintaining control of the launch programs.

In summary, during the last three decades, the
practice of both basic and applied science has become a larger and more institutionalized enterprise,
often organized around a set of large and expensive
research tools. The growth of scientific and technological information during the last three decades
has forced increased specialization and this has, in
turn, made the team approach mandatory in larger and
larger segments of science and technology. The ex-
panding efforts to apply science and technology to
the solution of society's problems have led to
broader and more vigorous programs of applied research, and this activity is particularly dependent
upon the formation of research teams, merging both
scientific and engineering talent. The broad profile and science and technolgy in 1979 is markedly
different from what it was in 1939, and it is likely
that the changes in the practice of science and
technology have been at least as great as any
changes in the public perceptions of the activity,
and perhaps even greater.

THE PUBLIC SUPPORT OF ORGANIZED SCIENCE

To support the growth of organized science, an increasing level of governmental -- primarily federal
-- support has been required. Prior to the Second
World War, governmental support for research was
focused almost exclusively on agriculture, with a
small portion assigned to medical and health research. Greenberg (1967) has appropriately termed
prewar American science an "orphan." Reagan (1969)
estimates that federal expenditures for all research
and development in 1940 -- the last prewar year --
were under $95 million, with the Department of Ag-
riculture accounting for about 30 percent of the
total.

As the nation mobilized for war, federal funds
became relatively plentiful for a variety of scientific and technological activities. Several new national laboratories were created, and substantial
resources were devoted to both capital and operating
expenditures. By the close of the Second World War,

the federal government had become the national patron of organized science and the dominant supporter
of basic research in the physical sciences (see Fig.
1.1).
     In the three decades since 1945, the federal
government has remained the principal financial supporter of organized science. In the basic physical
sciences and in the health sciences, the federal position has become so dominant that it approximates a
monopoly in numerous specialty areas.
     During the last decade, the rate of growth of
federal support stabilized and then declined slightly in constant dollar terms. Some disciplines and
fields have suffered relatively large declines in
constant dollars and some areas -- like energy-
related work -- have experienced continued growth.
While it was mathematically impossible for the rate
of growth of the first two postwar decades to continue, the impact of stabilizing federal support on
the organized science community has increased the
awareness of many scientists of the extent of the
federal role in this area.
     The reduction of the growth rate has had a
number of important effects on organized science.
Throughout the postwar years, the growth of organized science demanded an ever larger flow of skilled
manpower, especially doctorally-trained scientists.
In most university laboratories and research centers, the conduct of basic research has been built
on a large supply of advanced graduate and postdoctoral students who perform increasingly longer
apprenticeships and who expect, as a quid pro quo,
professional employment, typically in a research-
oriented university. Many scientists failed to
realize that these expectations could be met only by
a moderate to high growth rate, and the relatively
sudden drop to a non-growth posture in the early
1970s produced a small scientific unemployment crisis. While the actual unemployment rate of doctor-
ally trained scientists probably never exceeded 2
percent (NSB 1977), the shock, coming after three
decades of sustained growth, was sufficient to raise
the anxiety level of the scientific community markedly and to cause numerous universities to reassess
the number of students admitted to advanced programs
in the sciences.
     As the scientific community began to focus its
attention on the growth-rate question, many scientific leaders were surprised at the extent of the
dependence of organized science on federal support.

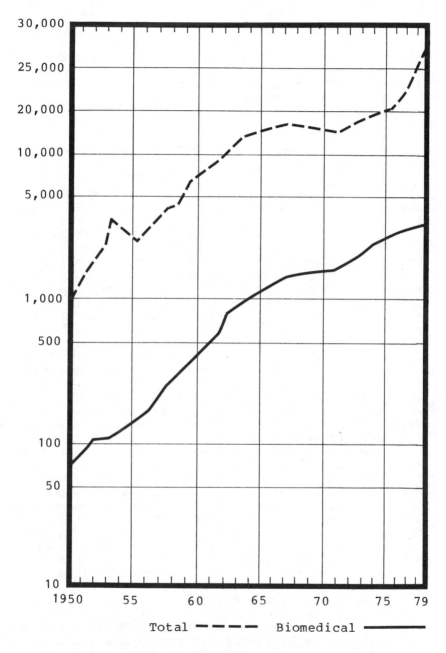

Fig. 1.1.   Federal obligations for research
          and development, 1950-78.

Source: American Association of Medical Colleges

As the level of support for organized science became
a contested issue between the Congress and Presidents Nixon and Ford, the full extent of the public
scrutiny and control of organized science became
increasingly clear. The belated discovery of this
trend was a surprise to both political leaders and
the scientific community.

THE PUBLIC REGULATION OF ORGANIZED SCIENCE

The mini-recession of the early 1970s in organized
science served notice of the potential impact of political decision making on the nature and operation
of organized science. For most of the first three
decades after 1945, organized science in the United
States operated within a structure of federal and
university funding and peer governance. While it
was technically necessary for the president to propose budgets and programs and for the Congress to
enact them, the actual operation of the major federal scientific agencies was in the hands of administrators selected with the advice and consent of the
scientific community, and the grant-making process
was dominated by panels of university-based scientists (Price 1965). The availability of resources
was ample, and, in the decade after Sputnik, resources were abundant. If this was public regulation, the leaders and the practitioners of organized
science were nearly unanimously in favor of it.
    It was only when the resource situation changed
from an N-sum game to a zero-sum game that the control aspects of the system became apparent. When
all areas of organized science were growing, the
priority implications of a higher growth rate in one
area than another was of little concern. After all,
the major problem for most laboratories was to re-
cruit, organize, and obligate the available funds
within each fiscal year. When addition turned into
subtraction and reallocation, the control inherent
in priority setting became clear.
    In retrospect, the federal role in priority set-
ting can be seen much more clearly than was possible
at the time. During the war years, the decisions to
initiate the Manhattan Project, to encourage the
development of numerous synthetic compounds, and to
emphasize aeronautical engineering were dictated by
strategic considerations involved in winning the
war, but each decision established major clusters of

organized scientific activity that continued to
influence the priority setting processes in the decades that followed. Would Oak Ridge or Argonne National Laboratories have been initiated and built in
the late 1940s or the pre-Sputnik 1950s? Probably
not.

In the postwar years, the growth of the National
Institutes of Health (NIH) shaped medical research
in the United States. The NIH emphasis on disease
and organ groupings tended to focus work on acute
care problems, although it was unquestionably successful in eliciting ever greater public financing
for the medical sciences. Through the informal
influence of the peer review system, a portion of
these resources was regularly used for basic work
not assignable to a disease or organ category, but
little was diverted for preventive or ecological
investigations. Undeniably, our present, very expensive acute-care medical system is a reflection,
in part, of this organ-disease emphasis.

Perhaps the clearest example of the federal priority setting is the manned space program. In a decision apparently initiated by speech writers rather
than representatives of organized science, President
Kennedy (with the consent and support of the Congress) launched the nation on the multibillion-
dollar Apollo Program. The magnitude of the program
utilized a significant portion of university and
industrial resources, shaped the research careers of
innumerable young investigators, and became a special field in engineering. It was, in Simon Ramo's
(1970) words, a "science olympics," and it was fun.
But it was a decision to use science resources for
one objective and, therefore, not to use them for
other objectives, at least directly. By the end of
the 1960s, the import of this decision was apparent
to the scientific community, the public, and the
Congress, and reductions in space-related expenditures were made.

The present emphasis on energy-related research
illustrates the same process. As the dimensions of
the energy problem became clear, the federal government began to allocate more and more resources for
energy-related research, and numerous laboratory directors and research scientists began to ponder the
possible energy-related aspects of their work. This
is not to suggest that scientists radically changed
their work, or changed fields, for this rarely happens. Rather, a physicist who is interested in a
topic like superconductivity, and who may have been

studying the process for several years before the
nation's new energy awareness, now sees a potential
application for his work, and he may be able to secure additional resources to expand and expedite his
work. This hypothetical investigator may have been
funded by the National Science Foundation or his
university in prior years and would have been supported in future years without an energy crisis, but
the relative priority placed on his work is now significantly higher, and the resources available to
him may be substantially greater. He will be able
to attract additional graduate and postdoctoral students to his project, expand his research team, and
execute his studies more rapidly. It is a subtle
form of control, mediated by panels of peers, but it
is a form of control or regulation nonetheless.
     The political process of priority setting is
perhaps the most important form of regulation of organized science, although the most recent to be recognized. The adoption of the new budget process in
the Congress and the establishment of an active Congressional Budget Office has significantly changed
the decision making process. In prior years, the
Congress debated and adopted each departmental budget separately or in related groups, so that support
for scientific research was usually reviewed on its
own virtues and generally fared very well. Now that
the Congress must adopt a total federal budget, it
must make comparative judgments. It is not a question of whether a proposed appropriation for scientific research is good or bad, but whether it takes
priority over medical care for the elderly, the construction of municipal sewage systems, or increased
revenue sharing. Once the Congress establishes its
annual budget ceiling, the process becomes a zero-
sum game for any given function, and the scientific
community is only beginning to understand the implications of the process (Roback 1977; Adams 1976).
     Ironically, a combination of other factors has
increased the centrality of federal financing of organized science just as the nature of the process is
changing. The majority of basic research in the
United States is conducted in university-based laboratories and centers, and the alternative funding
for organized research in both public and private
universities has stabilized or declined as much, if
not more, than federal support. The high unemployment and inflation rates that have characterized the
American economy in the 1970s have placed substantial pressure on state treasuries, and support for

state universities has generally lagged behind the
inflation rate, creating a loss of support in constant dollars. The depressed level of the stock
market during the same period has reduced the value
of the endowment or capital base of most private
universities and foundations, reducing internal
university budgets and foundations grants simultaneously. In sharp contrast to the relative benefi-
cence of state governments, of the stock market, and
of private foundations during the 1960s, laboratory
directors in the 1970s have faced hard choices.
Federal funds continue to be the dominant source of
support, and internal university or foundation alternatives are rarely available. Changes in federal
priorities now have a relatively larger impact.

During the last decade, the more traditional
forms of governmental regulation have increasingly
included organized science. Through a network of
laws and executive orders, recipients of federal
funds have become obligated to implement affirmative
action programs, to increase protection against occupational hazards, to modify buildings to serve the
handicapped, to protect human subjects in research,
to provide more humane treatment for animals used in
research, to follow specified accounting procedures,
and to provide environmental impact statements for
new structures. For the most part, these regulations were designed for all groups and firms doing
business with the federal government and were not
targeted at universities or other organized science
institutions. A few sets of regulations, especially
those dealing with human and animal subjects, were
designed primarily for organized science. Regardless of intent, practicing scientists and research
managers have experienced a substantial growth of
direct federal regulation in recent years, creating
a growing wariness of federal controls.

The recent confrontation of the Cambridge (Massachusetts) City Council and Harvard University and
the Massachusetts Institute of Technology over the
conduct of recombinant DNA research has further un-
settled the scientific community. While there is
undoubtedly a division of opinion within the scientific community over the relative risks of various
forms of genetic engineering, there is substantial
consensus that the issue ought not be decided by
local city councils. The question of who should decide this and similar disputes is a difficult issue
for the scientific community, and it is out of this

dilemma that proposals for a "Science Court" have
emerged.
     In summary, the growth of organized science in
recent decades has required substantial governmental
funding, and with the support has come a complex set
of direct and indirect controls. The pervasive influence of the federal government in priority setting has become increasingly apparent, although the
scientific community retains substantial influence
in the process. Direct federal regulation of organized science has increased in recent years, but it
has been largely procedural rather than substantive.
The intervention of a local government in a substantive issue in a showcase confrontation has raised
the anxiety level of the scientific community.

                  THE OUTLOOK FOR THE FUTURE

In general, it is reasonable to expect each of the
three trends discussed above to continue for the
balance of this century, although at somewhat different rates.
     The trend toward "organized science" and "big
science" is almost surely irreversible. In some
fields -- like subatomic physics or astronomy -- it
is virtually impossible to conduct research without
access to multimillion-dollar facilities. In numerous fields, the reliance on teams of specialists and
expensive equipment will dictate an increasing level
of organization. In a few fields -- like mathematics -- the availability of inexpensive microproces-
sors may reduce the dependence on central computing
systems and reduce the necessity for "organization."
On balance, however, it would appear that the degree
of "organization" will continue to grow.
     Similarly, the dependence of organized science
on federal support will certainly grow. Demographic
projections for the balance of the twentieth century
indicate declining university enrollments, placing
even greater pressures on institutional budgets.
Fewer university teaching positions will be available, and it is unlikely that state budgets or private endowments will sustain a large number of
positions exclusively for research. Industry has
shown limited interest in basic research in the past
and shows no special promise for the future. The
federal government will increasingly become the patron of organized science in the United States.

Despite the hopes of the scientific community, it is unlikely that the regulatory impact of governmental decision making will decrease. The competition for federal funds will continue to be intense and the inherent priority setting nature of budget making will remain. It may be possible to increase the participation of the scientific community in the decision making process, but it is more likely that there will be increasingly divergent views within organized science and that the adjudicating function will be more often in the hands of elected officials and their staff agencies.

If this assessment is correct and issues important to organized science will be in the public arena with increasing frequency, it is essential to understand the attitudes of the American public and their leaders toward organized science. It is that task to which we now turn.

2 The Public Understanding of Organized Science

In the last two decades the number of scientific and technological issues actively debated in the political arena has increased markedly. The congressional debates over the supersonic transport, the anti-ballistic missile system, the health hazards of cigarettes, the testing of new drugs, and the environmental impact of fluorocarbons illustrate the growing number of political issues that involve, in part, scientific judgments. Shen (1975) estimates that slightly over half of the bills introduced in the Congress concern a scientific or technological topic. While there is broad agreement with the idea that a democratically governed populace ought to be more knowledgeable about organized science, there is much less agreement about what constitutes an appropriate level of public understanding. Three major concepts have appeared in the literature and in policy discussions.

First, the term "understanding of science" is often used to refer to a rational and systematic approach to problem solving. This concept was particularly popular among elementary and secondary science educators during the 1930s (Davis 1935; Hoff 1936; Noll 1935) and has experienced a major revival during the last decade (Baumel and Berger 1965; Kahn 1962; Lowery 1966; Schwirian 1968; and Vitrogan 1969a). Etzioni and Nunn (1974) have suggested that Rokeach's dogmatism scale is inversely related to "scientific thinking" (Rokeach and Eglash 1956).

In a very fundamental sense, the principles of democratic government assume a large measure of rational judgment by citizens and, especially, the

electorate. If citizens and electors made judgments
and took actions on the basis of whimsy or irratio-
nal factors, there would be little hope for the survival of a democratic government in any nation. As
V. O. Key, Jr. argued in The Responsible Electorate
(1966):

> The perverse and unorthodox argument of this
> little book is that voters are not fools.
> To be sure, many individual voters act in
> odd ways indeed; yet in the large the elec-
> torate behaves about as rationally and
> responsibly as we should expect, given the
> clarity of the alternatives presented to it
> and the character of the information avail-
> able to it (p. 7).

Viewed in this context, rational thinking is much
broader than any concern with organized science and
not a useful definition for our purpose.
    Second, an "understanding of science" is often
taken to mean a command of current scientific terminology and theory. The best example of this approach is the National Educational Assessment
studies which have asked national samples of students and young adults to define or explain a radi-
us, a diameter, the difference between Fahrenheit
and Centigrade scales, and similar items (NAEP 1971,
1973). The majority of science-education works in
the last three decades have focused on this measure
of scientific literacy, although frequently stressing innovative means of transmitting the information
(Allen 1959; Aiken and Aiken 1969). Undeniably, it
is important to be able to understand the basic set
of terms in which scientific issues are discussed.
But once one moves beyond a minimal scientific vocabulary, there is much disagreement about the levels of substantive knowledge needed for responsible
citizenship, with scientific groups tending to set
higher standards and public-interest groups tending
to accept lower levels. Without attempting to re-
concile those differences, it would appear that an
"understanding of science" requires some level of
substantive science information or knowledge.
    Third, the concept "understanding of science"
is increasingly used to refer to the public's knowledge of, and attitudes toward, organized science.
The works of Davis (1958), Withey (1959), LaPorte
and Metlay (1975a, 1975b), Taviss (1972), and Miller
(1976) have focused on the public understanding of

the organized scientific enterprise as have the
first three Science Indicator reports by the National Science Board (NSB 1973, 1975, 1977). This
conception of "understanding" focuses on the comprehension of the structure of organized science, its
potential service and risk to our society, and its
political and social implications. In one sense,
this conception refers to "issue literacy" for
science-related issues, or what Shen (1975) has
called "civic science literacy." And, as the number
of science-related issues in the political arena
continues to grow, the importance of this facet of
the public understanding of science increases
accordingly.
     In summary, it appears that the public under-
standing of organized science is best defined as a
two-dimensional construct involving both some level
of substantive scientific information and some level
of science-issue literacy. In chapter four, we will
attempt to construct an operational definition of
these dimensions.

OPINIONS, ATTITUDES, AND IDEOLOGIES

As a preface to an examination of the literature
concerning the public understanding of organized
science, it is important to distinguish between
opinions, attitudes, and ideologies (sometimes termed belief systems). Hennessy (1972) has provided a
useful set of organizing definitions and concepts.
Hennessy uses "opinion" to refer to "immediate orientations toward contemporary controversial political objects" (p. 28). Opinions tend to be fragile
and lightly held. If an issue were raised in two
different situations by different persons, an opinion might change. In contrast, Hennessy defines an
"attitude" to be a more enduring and diffused orientation toward an object, including political issues
and topics not controversial or salient to most
adults. Attitudes are more deeply rooted, more
solidly supported by information and interest than
opinions, and less susceptible to fluctuation.
     The concept of a belief system or "ideology" is
rooted in the work of Converse (1964) and refers to
a network of attitudes that reflects either a logical consistency or a higher-order philosophy. A
classic example is Marxism, but the concept of
ideology is frequently applied to conservatism,

liberalism, militarism, pacificism, and similar
constellations of political interest. Persons with
political ideologies are usually very interested in
political matters, regular consumers of political
information, and likely to be active in electoral
and other policy-influencing activities.

The primary focus of this study will be public
attitudes toward organized science and the implications of those attitudes for science policy. On
low-saliency topics like organized science, it is
unlikely that most persons even have opinions (except when confronted with forced-choice questions)
and that the ephemeral nature of these opinions
provides little basis for an analysis of the public
understanding of science. By focusing on the presence or absence of attitudes toward organized science, and subsequently on the contrast of those
attitudes, we may be better able to comprehend public involvement in science policy. The relationship
of attitudes toward organized science and broader
ideologies is a topic of considerable interest, but
beyond the scope of this analysis.

GENERAL POLITICAL ATTITUDES

Since attitudes toward organized science are but one
relatively specific species of the genre of public
policy or political attitudes, it is important to
begin with a brief discussion of general political
attitudes. From Gosnell's pioneering work 50 years
ago to the numerous surveys of today, social scientists have been seeking to understand the development, structure, and modification of political
attitudes and activities, especially voting (Gosnell
and Merriam 1924; Campbell, Converse, and Stokes
1960; Milbrath 1965; Verba and Nie 1972). While the
remaining disagreements among social science researchers are important, it is possible to distill
from this growing literature the structure of a
general model that would command considerable consensus among social scientists.

As a starting point, it is necessary to recognize that political activities in general are not
salient to most adult Americans. A bare majority of
eligible American adults cast a ballot in presidential, senatorial, or congressional elections, and it
is a rare municipal election that draws half of the
eligible voters to the polls. In nonpresidential

years, almost all elections and referenda are decided by a minority of eligible adults.

Almost three decades of survey research documents the low levels of citizen interest in and knowledge of politics. Converse's seminal analysis (1964) of political attentiveness outlined a mass public with sketchy political information, low levels of political interest, and few well-developed political attitudes of any kind. Converse estimated that not more than 15 percent of the adult population could be classified as having an organized political-belief system by even relatively lenient definitions. Jennings and Niemi (1974) applied Converse's measure of belief systems to a national sample of high school seniors in 1965 and that found only 16 percent had organized political-belief systems, suggesting little generational improvement. Hennessy (1972) goes further:

> I submit...that (1) not more than 10 to 30 percent of the voting adults in America have even a rudimentary [political-belief system]; (2) that many people do not have political attitudes, for we have said that attitudes are tendencies -- tendencies that rest upon some sense of categorization and appropriateness that necessitates a minimum of integration of things political; and therefore (3) that political opinions may, and for some significant number of average people, do, exist as more or less unintegrated responses to stimuli of the moment (p. 35).

The works and conclusions of Converse, McClosky (1964) and others (Prothro and Grigg 1969; Field and Anderson 1969; Luttberg 1968) produced a generation of social-science research seeking to refute or significantly modify this relatively dismal portrait of participatory democracy in America. Nie and Anderson (1974) have examined more recent survey data and found higher correlations among attitudes on various policy clusters, and they have claimed a significant growth in the number of persons with belief systems. Sullivan, Piereson, and Marcus (1978) have demonstrated that most of the growth in attitudinal constraint reported in earlier studies can be explained by changes in question wording introduced by the Michigan Survey Research Center during the analysis period.

In summary, then, it is necessary to start with
a recognition that political activities in general
are of relatively low saliency to most Americans and
that specific policy attitudes are extremely rare.

THE ADULT UNDERSTANDING OF ORGANIZED SCIENCE

During the last two decades, there have been several
major studies of the institutions and programs most
closely associated with organized science. In general, the available data indicate a decline in the
level of unreserved esteem for science and technology and a growing awareness of the doubled-edged
nature of the enterprise, that is, its enormous promise for good and its equally significant potential
for harm.
     In 1957, the National Association of Science
Writers (NASW) sponsored a national survey concerning adult attitudes toward science and technology.
The survey (Davis 1958) was conducted just before
the first Sputnik launching and provides invaluable
baseline data. A second survey, conducted several
months after Sputnik assesses the impact of that
event. Interestingly, the Russian space launching
made no immediate difference in the general pattern
of adult attitudes toward science and technology
(Withey 1959).
     The NASW surveys found that the great majority
(92 percent) of Americans felt that science was making their lives "healthier, easier, and more comfortable" and that 83 percent believed that the
world was better off because of science. Eighty-
seven percent of the survey subjects indicated that
science "is the main reason for our rapid progress,"
and 64 percent rejected the idea that science breaks
down people's ideas of right and wrong. About 10
percent of the respondents in the NASW surveys indicated some reservations about the contributions of
science. Withey (1959) reported that:

     The contributions of science to the arma-
     ments of war were seen as the major area of
     bad effects of science. Most of the devel-
     opments in atomic science were seen as fall-
     ing into this area of threat. Also, most of
     the new advances in space were seen more in
     the area of an international race with Rus-

sia, and the total context of the "cold
war," than as an advance in science per se
(p. 387).

While only 10 percent doubted the net positive
benefit of science, there were indications of a
higher level of unrest. A full 47 percent felt that
science "makes our way of life change too fast," and
40 percent expressed concern that "the growth of
science means that a few people could control our
lives." It is interesting to note that the only set
of responses that changed significantly between the
first and second NASW surveys was an increase in the
percentage of respondents concerned about the con-
centration of societal control in a few hands. Although a majority pointed to positive contributions
from science in improved health, a higher standard
of living, and industrial and technological improvements, 49 percent rejected the idea that science
might be able to "solve social problems, like crime
and mental illness."
Withey's assessment of the public mood in 1957-
58 was that:

On the surface the natives are quiet, supportive, and appreciative but there is some
questioning, some alert watching, and considerable mistrust. The public will wait
and see; they have no reason to do anything
else, and many have no other place to turn
(p. 388).

In the late 1950s and early 1960s, several academic investigators studied community conflicts over
the fluoridation issue and noted that opponents of
fluoridation often held broader antiscience attitudes (M. Davis 1960; Kirscht and Knutson 1961;
Mausner and Mausner, 1955). Unfortunately, interest
in the roots of these antiscience attitudes appeared
to subside with the fluoridation issue and the literature contains no efforts to expand these findings
into more general models.
The public awareness of the potential negative
outcomes of modern science continued to grow in the
1960s. Oppenheim (1966) combined the 1957-58 SRC
data with a 1964 survey by the National Opinion Research Center (NORC) and reported a rising level of
public reservation about some of the ramifications
of science.

In 1972, the National Science Board (NSB) initiated a biannual survey to measure trends in the
attitudes of Americans toward science and technology. The 1972, 1974, and 1976 surveys represent very
extensive data sets, but the design of the studies
limit their utility. The NSB (1973, 1975, 1977)
surveys found that the public attitude continued to
be supportive, but less so than in 1957 and 1958.
Seventy percent of the respondents indicated that
they felt that science and technology had changed
life for the better, but only 54 percent were willing to say that on balance science and technology
had produced more good than harm. Slightly more
than half of the American people believed that the
pace of change resulting from science and technology
was "about right," while approximately 20 percent
felt that the pace was too fast. A survey of three
Massachusetts cities focused on technology and its
products and found that 76 percent of the respondents felt that technology does more good than harm
(Taviss 1972).
     Further, there was a more manifest public un-
derstanding that science and technology had provided
a mixture of desirable and undesirable outcomes. On
the basis of a comprehensive analysis of public support for the opposition to several science-related
issues, Louis Harris concludes that public wariness
toward science has been building primarily because
of potential environmental damage from some scientific activities and the associated threat to human
health (Harris 1973). On the question of the net
benefit of science and technology to society, 31
percent of the public indicated that the positive
and negative results of science were about equal.
More than half of the respondents indicated that
"some" or "most" of our problems were caused by
science and technology. Despite the increasing
awareness of some negative outcomes, a substantial
plurality of the public indicated that the degree of
control of science and technology by society should
stay the same (NSB 1973, 1975, 1977). A 1972 Harris
Survey found essentially the same pattern.
     It is important to understand that the growing
public wariness of the potential negative outcomes
from science does not represent an antiscience or
antitechnology outlook. In almost every survey, respondents continued to express high levels of expectation for good outcomes from science at the same
time that they expressed some wariness of the negative aspects. It would be prudent, however, to

recognize this growing public wariness as the raw
material for a more overt antiscience attitude,
should there be major harmful outcomes from scientific or technological activities in the future.

On the other side of the coin, the public expectations of science and technology appear to be
higher than the earlier NASW study. Whereas 49
percent of the American public doubted that science
could solve societal problems like crime and mental
illness in the 1957-58 surveys, by 1972-74 a full
three-quarters of the public felt that science and
technology would eventually solve some or most
problems like "pollution, disease, drug abuse, and
crime."   When asked to select the areas in which
they would most like to have their taxes spent for
science and technology, the public's preferences
were health care (60 percent), reducing and controlling pollution (50 percent), reducing crime (58
percent), finding new methods for preventing and
treating drug addiction (48 percent), and improving
education (48 percent). The areas in which the public was least supportive were space exploration (37
percent) and developing or improving weapons for
national defense (30 percent) (NSB 1975). In a 1972
study of attitudes toward science and technology
among California adults, LaPorte and Metlay (1975a)
found similar patterns of public support for science
and an appreciation of the dual potential for good
and harm.

THE PREADULT UNDERSTANDING OF ORGANIZED SCIENCE

In sharp contrast to the numerous studies of child-
hood and adolescent attitudes toward science courses
and scientific careers, there has been little attention devoted to what Shen (1975) has termed "civic
science literacy." There has also been a significant level of effort devoted to the study of "scientific thinking," which most often reflects a level
of logic that ought to be expected of all citizens,
regardless of their particular interest in science
per se.

The only overtly methodological study is the
work of Roshal, Frieze, and Wood (1971). Starting
from the premise that children should "appreciate
the benefits of technology and...not be afraid of
the many machines in their environments," the investigators constructed a 24-item scale designed to

measure attitudes toward "personal control and understanding of machines, man's ability to control technology, and the positive benefits of technology." The Attitude Toward Technology Scale (ATT) was tested with 610 sixth-grade children and produced an average score of 3.2 on a scale from one (very unfavorable) to five (very favorable), a modest level of support at best.

As one part of a broader study, Lowery (1967) studied the images of science held by a group of 335 fifth-grade students and found a number of modal images. He noted that a significant number of children held images of science relating to space exploration, that most children's images related to the products of science rather than its processes or principles, and that there was no understanding of the difference between science and technology.

By the time young people reach high school, they have a much more comprehensive view of science and technology; there is the emergence of an understanding that it involves the potential for both great benefit and immense harm to mankind. For more than two decades, the Purdue Opinion Panel has regularly surveyed a nationwide sample of approximately 3,000 students, and the questions relating to science and technology paint a portrait of broad support for science and technology but an awareness of potential and present problems. Eighty percent agreed that the goal of science "is to benefit mankind," and 77 percent said that were it not for science and technology "we would be living in ignorance and disease" (Remmers 1957).

While there is broad support for science and technology, most high school students do not see these activities as a primary source of national growth and improvement (Remmers 1957). When asked to select among several societal institutions those "most important to maintain our greatness," high school students ranked science last:

        Education 51%
        Religion 30%
        Civil Rights 11%
        Science 7%

Even on a question about the factors responsible for the American standard of living, only 9 percent of high school students cited science and technology as "the most important reason":

```
Our democratic form of government . . . . . 29%
Our belief in the freedom of all men . . . 22%
Our free educational system for all . . . . 17%
Capitalism and profit-making in business  . 14%
Our science and inventions . . . . . . .  9%
Other . . . . . . . . . . . . . . . . . . 7%
```

As a part of a study of attitudes toward science as a career choice, Allen (1959) found that a sample of 3,000 New Jersey high school students held very positive attitudes toward science and technology. Eighty percent of the students felt that "science and technology are essential to the development of present day cultures," while only 6 percent would agree that "science and its inventions have caused more harm than good."

The students studied by Allen realized that the pace of change in society was being accelerated by science and technology, but did not find this a major concern. Over 77 percent of these high school students agreed that science and technology are "primarily responsible for the frequent changes which occur in our manner of living." Almost 65 percent of these respondents felt that "decisive economic, political, and social processes are greatly influenced by science." At the same time, 89 percent of the students rejected the idea that "science has done little for the average citizen."

A QUESTION OF SALIENCY

In the literature reviewed above, the general assumption is that public policy issues concerning science and technology are the business of all citizens, and the yardsticks of support have been applied to general population samples. Most investigators have drawn the reluctant conclusion that science information is indeed complex and that the educational level of the general population prevents many people from truly understanding the scientific issues of the day and from enjoying the beauty and elegance of scientific theory. Lord Snow and others have suggested that there should be greater scientific understanding among nonscientist intellectuals and greater nonscientific understanding among scientists, but few have dared to suggest that everyone will really understand science in this generation. We are left with the inevitable conclusion that a long-term solution demands more attention for science education. This logic, however, provides no short-term policy guidance.

The problem of a low level of public information or concern about organized science is not unique, and it is at this point that the broader social science literature becomes especially helpful. For any given person, the range of possible areas of interest is vast. One of the characteristics of modern society is that the volume of information is overwhelming and no single individual can become knowledgeable or remain current in more than a relatively narrow range of topics. Ironically, many groups of persons with an intense interest in an area tend to think that everyone ought to be knowledgeable about that particular area. The public apathy toward agricultural policy, foreign policy, the problems of the aged, prison reform, national health insurance, endangered species, environmental quality, and modern child-raising theory is decried regularly. In marked contrast to our frontier ancestors who waited eagerly for old newspapers from the East, the plight of modern man is to sample selectively from an avalanche of available information.

The problem of the public understanding of science and technology is not fundamentally different from the problem of low levels of public understanding of a number of important public issues. For example, only a small proportion of the adult population follows foreign policy developments among the nations of the world and fewer yet understand the symbolic exchange of words and actions that leads to the formulation of foreign policy in the twentieth century, yet the United States remains a healthy democratic society.

The parallel between science policy and foreign policy is both interesting and appropriate for our purposes. Both science and foreign policy are complex topics that require a good deal of background knowledge to follow on a day-to-day basis, and both have enormous import to the welfare of the American people. In general, the American people have paid little attention to and hold little specific information about either area. In both areas, there are well-developed specialized information channels, and small segments of the population are very knowledgeable about each area.

For our purposes, then, there are two fundamental questions: Why is organized science a salient

subject to so few Americans? How does this interest
in organized science develop? Fortunately, there is
a rich literature concerning the socialization of
preadults and young adults to politics and political
issues, and it is to this literature we now turn.

3 The Development of Issue Attentiveness

As public policy issues grow broader and more complex, the pressure for issue specialization increases. Given the evolution of organized science and its frequent involvement in public policy issues, it is important to examine the processes by which new entrants into the American political system acquire a basic interest in political affairs and by which issue attentiveness develops. This chapter will look first at the existing literature concerning the development of general political interests, then focus on the more general process of issue attentiveness, and close with a discussion of attentiveness to organized science as a specific case of specialized issue attentiveness.

THE DEVELOPMENT OF GENERAL POLITICAL INTEREST

In a modern democracy, the modal citizen may hold a variety of attitudes toward the government and the authorities who hold office in the government. At the ends of the spectrum, some citizens may be almost oblivious to the existence and policies of the government on a day-to-day basis, while other citizens may hold an extensive and well-organized set of policy views and evaluative attitudes toward the government. Easton (1953, 1965a, 1965b) has provided a very useful theoretical framework for thinking about the organization of attitudes toward the political system, and it is useful to turn briefly to

this conceptualization prior to engaging in a discussion of the content of the attitudes per se.
     Easton suggests that the political system can
be seen as an entity capable of making authoritative
allocations of valued outcomes, in contrast to voluntary or social groups that lack the ability to
enforce their decisions. The day-to-day policy and
program-execution decisions are termed outputs in
the Easton model, and the cumulation of a series of
outputs is termed an outcome. Citizens make demands
for various types of outputs and outcomes and offer
their support to the political system. Easton identifies two types of support -- specific support for
particular outputs or decisions and diffuse support
for the continuation and operation of the political
system itself. Without a reasonable level of diffuse support, it would not be possible for a political system to sustain itself for any extended
period of time without depending on sheer force.
Easton's theoretical framework is very useful in
understanding the rise and demise of political
systems.
     For the purposes of this discussion, the dis-
tinction between diffuse and specific support is a
useful construct. Too often, readers of opinion
polls interpret reports of declining support for the
president or for some specific policy as indicators
of a loss of confidence in the government in general. Editorial writers often refer to the erosion of
public confidence in a specific official or a specific policy as if it were an erosion of faith in
the basic institutions of government. It is important to distinguish between types of support and to
understand the relationship between specific and
diffuse support attitudes, since they are not independent in the long run. Since this analysis
focuses on issue attentiveness, it is especially
important to make this differentiation clear and to
set the proper framework for the analytic discussions that follow.

                The Development of Diffuse Support

During the last three decades, a substantial amount
of research has focused on the development of general attitudes toward the political system, and it
is now possible to describe the process of diffuse-
support development for the modal group of young

Americans in terms that would gain the agreement of
the vast majority of the social scientists working
in the field. In several early studies of preadult
political attitudes, Easton and Hess (1961, 1962)
and Greenstein (1960, 1965) noted the highly positive affective views of the president held by most
youngsters in the elementary school years. An
overwhelming proportion of the second-, third-, and
fourth-graders know the name of the president and
understand that he is an important public figure.
Hess and Torney (1967) suggest that this positive
affective orientation is similar to the feelings of
the child toward his own family and that this positive family affect is transferred in part to the
president as a personal figure and then to the institutions of the presidency and the government.
Hess and Torney observe:

> It is difficult for a child to comprehend a
> complex political institution. Although
> there are symbols of Congress, the Supreme
> Court, and other institutions of government,
> these are not used with the same frequency
> and ritual as the flag and the pledge of
> allegiance are used as symbols of the na-
> tion. It seems likely that complex social
> and political systems are initially concep-
> tualized as persons to whom the child can
> relate. Through attachments to these per-
> sons, the individual becomes related to the
> system. In short, to the child, the govern-
> ment is a man who lives in Washington, while
> the Congress is a lot of men who help the
> president (p. 39).

To young children, the president is a person who is
concerned about their ideas and who would help them
if they were to ask him. Hess and Torney report
that about 75 percent of second-graders expressed
the belief that the president would "care a lot"
about their ideas if they were to write him, and
that about 43 percent of eighth graders still held
this personalistic image of the president.
    As the preadult grows older, he begins to un-
derstand more about the complex nature of government
and is increasingly able to distinguish between the
occupant of an office and the institution or authority represented by the office itself. During the
elementary school years, the young child acquires a
much richer cognitive base and is able to provide

more detail in comparing the powers and functions of
the officeholders in the political system. Dawson,
Prewitt, and Dawson (1977) describe this growth
period:

> By age ten or eleven children begin to move
> away from the highly personal and emotional
> perceptions and to comprehend more abstract
> ideas and relationships. Where younger
> children can do little more than identify
> political leaders, especially the president,
> as powerful and benevolent, older children
> show a greater capacity to understand and
> identify certain tasks that go along with
> particular political roles. During this
> period, information and congitive content
> are added to the vague feelings and identi-
> fications acquired earlier (p.54).

By the end of adolescence, the modal young person is
much more at ease in dealing with abstractions and
in relating abstract concepts to the political and
social issues in his or her community. Concepts
like authority, rights, liberty, equality, and similar terms become comprehensible and manageable in
thinking about political affairs. The adolescent
also begins to develop an expanded time-perspective
by age 15 or 16, allowing both a long historical
look back and a longer perspective on the future.
The period is also characterized by an increase in
interest in politics for many young people (Dawson,
Prewitt, Dawson 1977; Adelson and O'Neil 1966;
Adelson 1971).

The preceding discussion has focused on the
modal group of young Americans, but there is evidence that this general model may not be applicable
to all segments of American society. Jaros, Hirsch,
and Fleron (1968) studied a group of elementary
school children in Knox County, Kentucky, and found
a pervasive cynicism about the president and the
political system. They conclude:

> Children in the relatively poor, rural Appa-
> lachian region of the United States are
> dramatically less favorably inclined toward
> political objects than are their counter-
> parts in other portions of the nation.
> Moreover, the image which these children
> have does not appear to develop with age in
> the fashion observed for others; there is no

indication that. a process conducive to the
development of political support is operative in Appalachia (pp. 574-5).

Jaros, Hirsch, and Fleron also note that children in
Knox County appear to have less positive images of
their own families and that there may be a transmission of negative political images from the family,
directly or indirectly. Interestingly, the children
with the most positive images of government and politics in the Knox County sample were children from
homes without a father present, suggesting that the
presence of a father was associated with less positive images of the political system.

In similar studies, Greenberg (1970) and Garcia
(1973) have suggested that similar deviations from
the modal model occur for Black and Latin youngsters, respectively. Hess and Torney (1967) and
Weissberg (1974) have reported results indicating
that children from lower socioeconomic class backgrounds are slightly slower in the developmental
process, although ultimately follow a pattern similar to the modal model.

Returning to the modal model, a good deal of
work has focused on the sources of political learning or influence. In the absence of any evidence of
a genetic transfer of political attitudes, it must
be assumed that the young person enters the world
with no political ideas and that these attitudes are
learned in some manner. The debate has focused primarily on the relative importance of the home and
school environments as sources of political learning. More recently, there has been some attention to
the role of media, especially television, in the
political learning process.

In the traditional view of the socialization
process, the central role has been assigned to the
family (Berger and Luckmann 1966). The primary socialization of the parent or significant other con-
veys a wide range of basic values and perceptions
about the relationships among people and the acceptable range of behaviors. These underlying personality traits may have long-term political consequences
(Sniderman 1978). Hirsch (1971) summarizes a widely
held position concerning the role of the family in
political socialization:

Most scholars of political socialization
agree that the family is one of the most
pervasive agents of political socialization.

This is to be expected, for the child is
usually exposed to familial influence for a
long period of time. Studies have demonstrated...that the child tends to assimi-
late the party preference of the parent and
to look to the parent as a source of political advice. However, as the child grows
older and as contact with other agents increases, the parents' influence appears to
decline (p. 33).

In a national study of high school seniors,
Jennings and Niemi (1974) found considerable
variance in the congruence of parent and child attitudes. They observe that one of the major determinants of the success of parental transmission of
attitudes is the degree of homogeneity between the
parents and the relationship of the parental views
to those of the community, including teachers and
other norm-setting influences. On balance, Jennings
and Niemi conclude that the family can be an effective transmitter of political values under certain
conditions, and very ineffective under other conditions.

Hess and Torney (1967) contend that the family's primary effect is to "support consensually held
values rather than to inculcate idiosyncratic attitudes" (p. 113). Since the positive support
attitudes that characterize diffuse support are consensual in most political systems, it would appear
that the family has a substantial role in this area.

Hess and Torney (1967) see the school as the
most important single source of influence, asserting
that:

The public school appears to be the most
important and effective instrument of political socialization in the United States. It
reinforces other community institutions and
contributes a cognitive dimension to political involvement. As an agent of socialization it operates through classroom
instruction and ceremonies (pp. 120-1).

Hess and Torney suggest that the schools play a major role in the development of loyalty to the nation
and in the acquisition of the idea of citizen duties. The rules and procedures of the school are
seen as a learning procedure for future political

participation and for future acceptance of rule making by the political system.

 In summary, the literature concerning the development of diffuse support attitudes suggests that
these attitudes develop early, reflecting in most
youngsters a personalistic image of the government.
As a symbol of the government, the president is seen
as a person directly accessible to young children
and concerned about their needs. This original personal affective relationship is generalized to the
institutions of government as the young person ma-
tures and acquires a higher level of cognitive information about the political system. It appears
that the family is a major source of diffuse support
attitudes and that school plays a substantial re-
inforcing role, stressing patriotism and rule
obedience, and conveying a wide range of cognitive
information about the political system.

The Development of Specific Support

Specific support refers to those positive or negative attitudes that a citizen may have toward a
specific office holder or policy position of the
incumbent government. These attitudes represent the
individual's short-term evaluative position toward
the government; they are the grist of day-to-day
politics. It is Easton's position that the government's short-term popularity may ebb and flow in a
participatory political system, but that it is the
foundation of a strong level of diffuse support that
allows the short-term fluctuations in specific
support.

 For the purposes of this discussion, two types
of specific support will be identified and the development of those two types of attitudes will be
examined during the preadult and young adult years.
The first type of specific support to be discussed
will be the identification of an individual with a
political party. The second type of specific support attitude will be attitudes toward single-policy
decisions or positions. While these two types of
attitudes do not exhaust the full range of specific
support attitudes, they will serve to illustrate
certain modal patterns that typify most of the other
forms of specific support attitudes.

 The development of political party loyalties is
among the most extensively studied topics in Ameri-

can political science and there is a vast and still
growing literature concerning the processes and
causes of political party identification. Greenstein (1965) and Hess and Torney (1967) report that
children in the early elementary grades can report
which political party they personally favor, although they are unable, upon probing, to provide
any cognitive information distinguishing the two
parties. Greenstein reports:

> The prevalence of party identifications
> among New Haven children cannot be attrib-
> uted to 'response errors,' such as guessing,
> or the arbitrary checking of alternatives on
> the questionnaire. Children's party prefer-
> ences correlate appropriately with the demo-
> graphic patterns of partisanship in New
> Haven. They also are positively associated
> with favorable evaluations of the leaders of
> the same parties.
>
> The source of these preferences is, as often
> has been noted in the voting literature, the
> family. Only a handful of children in the
> entire sample indicated that their own party
> preferences differed from those of their
> parents. In interviews children explicitly
> speak of party as an attribute of the fami-
> ly. (Judith says, 'All I know is we're not
> Republicans.') Party identifications prob-
> ably develop without much explicit teaching
> on the part of parents, more or less in the
> form of a gradual awareness by the child of
> something which is a part of him. The pro-
> cess doubtless is similar to the development
> of ethnic and religious identifications (pp.
> 72-3).

Hess and Torney found a similar pattern of party identification in their national study of elementary school children, but they also observe that
most of the schools tended to teach the values of
political independence in evaluating candidates and
that the proportion of students reporting a strong
commitment to either one of the two major parties
declines with grade in school. This pattern would
suggest that the school socialization partially
erodes the family based party identification found
in the earlier years of elementary school. Following the same trend, Jennings and Niemi (1974) report

a significantly higher proportion of independents
among high school seniors than is found among their
parents.

In general terms, party loyalty might be seen
as a form of commitment to a set of issue positions,
or to a set of policies and positions that benefit
some particular group. Since political parties in
the United States have been notable for their lack
of cohesive ideology, one must be careful not to
overemphasize the group policy aspect of parties.
At the same time, Greenstein (1965) and Hess and
Torney (1967) report that most of the youngsters who
professed to have a party identification viewed the
identification in group support terms, and Converse
(1964) reports that a significant proportion of
adults conceptualize parties in the same manner.

Following this logic, it is reasonable to expect that persons would be more likely to give
short-term specific support to the officeholders and
policies of their own party and that they would be
more critical of the policies of incumbents from
other political parties. This is a useful illustra-
tion of the interaction between diffuse and specific
support. When the Democrats or Republicans lose a
particular election, they surrender the office in
question and those persons who voted for the losing
candidate typically wait for the next election to
try either to gain more support for the defeated
candidate or to field a new candidate that may be
able to draw sufficient support to win. It is the
existence of a high level of diffuse support for the
political system that allows the partisans of the
party out of power to sustain its commitment to the
political system and to wait for another election
and another try for control of the various offices.

In this context, specific issue orientations
can take on several different meanings, depending on
the level of partisanship involved. For strong partisans who are committed to party above all else,
it is unlikely that day-to-day issue debates will
change their positions on their support of specific
candidates or parties, and it is likely that they
will use their partisanship as a cognitive screen
for receiving and assessing issue information. At
the other end of the spectrum, the classic independent eschews any party loyalty and seeks to examine
each candidate or issue on its own merits and without a prior ideological commitment. For this type
of person, issues and the debates over issues are
the grist for political commitments and actions, and

the orientations of an individual to a wide spectrum
of issues may or may not reflect a consistent political perspective. Converse (1964) suggests that
only a very small proportion of adults have a consistent liberal or conservative position over a
range of issues.

It is possible that any individual's partisan
loyalty and issue preferences may come into conflict
from time to time. In a study of the 1960 and 1964
presidential elections, Pool, Abelson, and Popkin
(1964) operationalized this conflict and termed it a
"cross-pressure." They were able to identify critical sets of voters in each election that were cross-
pressured and to use these groups to effectively
predict the outcome of the elections. For example,
in the 1960 election, Pool, Abelson, and Popkin predicted that Republican Catholics and certain southern and border state Protestant Democrats would be
significantly cross-pressured. By estimating various defection rates and by predicting a higher level
of nonvoting by cross-pressured persons, they were
able to make a very accurate prediction of the 1960
voting well before the actual election. Similarly,
before the 1964 election, they found that certain
older Republicans who were very favorable to Social
Security programs, especially Medicare, would be
cross-pressured by the Goldwater opposition to some
of those programs.

Greenstein (1965) and Hess and Torney (1967)
report that specific issue orientations were much
later in developing than were party identifications.
By the time the young person is in high school, he
or she typically has a wide range of specific policy
attitudes, especially on topics relevant to younger
persons (Jennings and Niemi 1974). Not surprisingly, there are significant differences between the
policy positions of many of the high school students
and their parents on selected issues. This growth
of personal policy perspectives is parallel with a
decline in partisanship and is a predictable result
of the independence bias of the schools noted by
Hess and Torney.

Tedin (1974) has studied the transmission of
specific policy views from parent to child and concluded that parents are able to effectively transmit
their policy preferences, provided that the issue is
highly salient to the parents and that the child
correctly perceives the parental position. In most
cases, if an issue is highly salient to the parents,
they will be more likely to discuss it in the pres-

ence of the child, and this process may serve to
transmit a correct image of both the parental position on the issue and the high level of importance
attached to the issue. Jennings and Niemi (1974)
observe that the transmission of specific policy
views is enhanced when both parents hold a similar
issue position, thus conveying a uniform position to
the child.

In summary, the literature concerning the
development of specific support attitudes would
suggest that the preadult first develops a basic
identification with a political party, but an identification largely devoid of cognitive content in
issue terms. During the elementary school years,
the level of partisan commitment declines and the
level of cognitive issue information increases. By
the time a young person reaches the high school
years, it is not unusual for substantial issue differences to exist between parent and child. It
would appear, however, that parents can successfully
transmit policy positions on issues that they hold
to be important and about which they communicate
clearly and openly with their children.

THE DEVELOPMENT OF ISSUE ATTENTIVENESS

In the preceding chapter, it was suggested that the
vast range of active political issues exceeded the
ability of any individual to follow more than a
small subset of issues, a process that may be termed
issue specialization. While the causes and systemic
consequences of issue specialization are the topics
of other analyses, it is useful at this point to
note briefly two of the factors that appear to be
responsible for the accelerating pace of issue
specialization in the American political system.

First, there is an increasing premium on time
in all modern societies. While the hours of employment related work are declining in most industrial-
ized societies, the growth of social, recreational,
and entertainment activities has increased markedly
and these are strong competitors for each individual's time. Political activities are only one of a
wide range of alternative pursuits, and the recent
work of Verba and Nie (1972) would suggest that interest in political events attracts a declining
proportion of the adult population. Similarly, an
analysis of voting participation during the last

three decades displays a steady decline in all
elections and a strong decline in nonpresidential
elections at the state and local levels. If time is
conceptualized as a zero-sum situation, it would
appear that the increasing pressure on relatively
fixed time resources mandates some degree of issue
specialization for most adults.

Second, in a parallel manner, it would appear
that the information threshold for being informed
about most major political issues is increasing. As
issues become more complex and as the number and di-
versity of information sources grow, the concerned
citizen must devote an increasing level of time and
effort to maintaining a reasonable familiarity with
any given topic. Consider, for example, the current
political debate on inflation, which would be considered a major political issue by a substantial
portion of the American people. One set of arguments concerns the price of imported oil and the
position of the dollar in the international monetary
system. To evaluate these positions, a citizen
would need to know something about the international
monetary system, the nature of the gold markets, and
the present and prospective strengths of several
other major national currencies. Further, it would
be useful to know something about the portion of
energy costs determined by the price of crude oil,
the refining costs of various types of crude, and
the pricing and markup policies of the major oil
companies. Clearly, following the debate about the
impact of OPEC price increases on inflation in the
American economy is no small task. It is possible
to listen to the short segments on the evening television news or to read summaries in a daily newspaper or weekly newsmagazine, but most people would
soon realize the near impossibility of deciding
between the arguments of the experts on both sides
without substantially more information than most
citizens have or regularly obtain. The experience
of being partially informed about an issue may
contribute to increased citizen frustration about
the feasibility of meaningful citizen input to many
public policy questions.

When the increasing premium on time and the
growing information threshold for many political issues are considered together, it would appear that
the imperative for issue specialization is strong
and will continue to grow. As the complexity of
issues increases and as the volume of information

grows, the modal citizen will be able to follow a
narrower and narrower range of issues.

The import of this pattern for the operation of
the political system is substantial. While a full
discussion of those issues is beyond the scope of
the present analysis, it is useful to note two of
the more prominent manifestations of issue specialization in the political system: legislative
specialization and the sources of campaign contributions.

Since legislative bodies are charged with the
responsibility of handling the full range of political issues, the imperative for specialization might
be expected to be felt first in this arena. The
committee structure of legislative bodies is a manifestation of the need for specialized expertise, and
most congressional committees have developed professional staffs of experts in the subject areas
covered by the committee. Increasingly, members of
legislative bodies view themselves as issue specialists and occasionally describe themselves in those
terms. Senator Fulbright was viewed by his colleagues as a foreign policy specialist, and it
appears that he shared that view. Similarly, Con-
gressman Mills was viewed as a tax expert and
Senator Ervin was widely recognized in the Senate
for his expertise on constitutional issues.

At the other end of the process, Common Cause
-- an organization designed to protect the "common
good" from specialized interests -- reported that in
1978 congressional candidates received more campaign
funds from organized interest groups than from individuals for the first time in American history.
This pattern may represent an increased awareness of
the importance of electoral outcomes by interest
groups, an increased level of resources available to
interest groups, a declining level of individual interest in electoral outcomes, or some combination of
these factors. In any case, it would appear that
the funneling of campaign funds through interest
groups will have the longer-term effect of encouraging issue specialization.

As the complexity of the issues in the political arena grows, fewer and fewer issues become the
focus of major electoral contests. Since political
campaigns, and especially those at national or
statewide levels, depend on communicating short and
effective messages through the media, it is imperative that the issues that become the cleavage points
for a contest be few in number and comprehensible to

a large section of the electorate. Inevitably, giv-
en those criteria, the personalities and group alli-
ances of the candidates become the major foci of
most campaign activities. It is a rare electoral
contest that provides the winner a clear position
mandate on more than a handful of specific issues.
The other issues are decided by the officeholder on
the basis of his personal views and the arguments
presented to him by special interest groups. Rosenau (1974) has referred to the process of seeking to
influence legislators and administrators on specific
issues as "citizenship between elections."

    The concept of citizenship between elections is
particularly important to the process of issue specialization. Since most authorities enter office
without any particular electoral mandate on a wide
range of issues of interest to limited groups within
the population, the process of organizing these
smaller populations of interested citizens and con-
verting their interest into overt political actions
is an important component of the policy formulation
process in modern political systems. In a seminal
analysis of this process in the area of foreign
policy, Almond (1950) provided a model of attitude
transmission into the policy process that is essential to an understanding of the process of issue
specialization, and this model will become the basis
of much of the analysis in the following sections of
this monograph.

The Almond Model

In his 1950 study of the public understanding of and
attitudes toward foreign policy, Almond outlined
four levels, or strata, of public opinion. The base
of his public opinion pyramid is the "mass public,"
or those persons not able or willing to devote the
time and resources necessary for becoming and remaining informed in a given policy or issue area
(see Fig. 3.1). It is important to understand that
almost everyone is a part of the mass public for a
large number of issue areas.

    Almond labeled those persons who were willing
and able to inform themselves in a specialized area
the "attentive public." Rosenau (1961, 1963), Hero
(1959, 1960), Cohen (1973), and Mueller (1973) have
operationalized the concept and provided a substantial empirical description of the attentive public

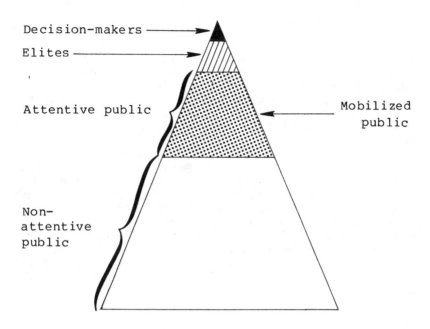

Fig. 3.1.   The Almond model of public attitudes.

for foreign policy. Presumably, there are also
attentive publics for science policy, agricultural
policy, civil rights policy, and numerous other
specialized interests. Given the investment of resources necessary to become and remain current in
a field, it is unlikely that any single individual
would be a member of very many attentive publics.
     Within attentive publics, there are variations
in the levels of knowledgeability and substantive
outlook among the persons comprising that group.
The opinion leaders of an attentive public are termed the "elites," and Almond focused considerable
attention on foreign policy elites. A number of
subsequent studies have provided a rich literature
on the elites in foreign policy (see Hero 1960;
Cohen 1959; Rosenau 1963) and in a limited number of
other policy areas (see Cobb and Elder 1972; Marwich
1968; Boynton, Patterson and Hedlund 1969).

Rosenau (1974) has divided the attentive public into a "mobilized public" and a nonmobilized segment. The mobilized public goes beyond interest and information acquisition and seeks to influence policy through overt actions. It is this mobilized group that writes legislators and executives, testifies at hearings, and privately lobbies decision makers to secure desired policy outcomes. Since the attentive public for most low-saliency policy areas is small, the mobilized public may be termed very small.

At the top of Almond's public opinion pyramid sit the decision makers. In foreign policy, science policy, and similar areas, the decision makers would include the leaders of the executive and legislative branches of the federal government, and perhaps the leaders of selected multinational corporations. Decision makers tend to be recruited almost exclusively from the policy elites in a given area. For those policy areas involving the federal government, there will be some overlapping of decision makers, but even here there is a relatively high degree of specialization (Price 1972; Fenno 1976; Orfield 1975).

In this model, the primary focus of policy debate is among the elites, with the attentive public serving as a sounding board. Since the elites in any given policy area may be expected to have some differences, the competing elites seek to mobilize those members of the attentive public who share their views to communicate their position to legislators and other authorities. Rosenau (1974) has studied this mobilization process in regard to the nuclear test-ban treaty and the civil rights voting bill in 1964, and his findings are consistent with Almond's (1950) description of the role of the attentive public and the elites in foreign policy formation.

ATTENTIVENESS TO ORGANIZED SCIENCE

In view of the utility of the Almond model in understanding a wide range of other low-saliency topics, it is appropriate to turn to the process of adapting the model to those issues concerning organized science. As a first step, it will be helpful to conceptualize the probable membership of each of the strata in the Almond model.

For the most part, the definition of the four strata will closely parallel those for foreign policy. The decision makers for organized science would include the president and his principal advisors and cabinet officers dealing with science-related topics; the leadership of the Congress, and especially the committee leadership of science-relevant committees in the House and Senate; the leadership of major scientific research centers and corporate research enterprises; and, occasionally, judicial officers, state officers, and local officials (Rettig 1977; Jachim 1975; Greenberg 1967). These persons have, as one of their powers, the authority to make binding decisions about the policy directions to be followed in various aspects of science. It is important to differentiate between policy decisions concerning organized science and the work of individual scientists (singly and in teams) actually engaged in scientific research. The former decisions are highly centralized in the United States, while the latter decisions are highly decentralized.

The policy elites in organized science are numerous and reflect the decentralized nature of the scientific enterprise in the United States and most Western European polyarchies. The elites are active in formulating policy alternatives and advocating particular solutions or programs. The science policy elite would include the leaders of the numerous scientific societies and disciplinary organizations, the directors of major research centers and laboratories, the top management of corporations active in organized science, editors and writers, and self-selected public interest advocates (Zuckerman 1977; Mulkay 1976; Whitley 1976; Nelkin 1977). For this latter group self-selected is not meant in a derogatory sense, but rather to denote persons who develop an interest in science issues, become knowledgeable, and exert influence and leadership among the attentive public. Rachel Carson would be an example.

The attentive public for organized science would include those persons who choose to inform themselves on scientific issues and topics and to follow the debate of issues by the elites. While a more formal operational definition will be offered below, the attentive public for organized science might be expected to include a significant number of practicing scientists, science-related businessmen, clergy, and citizens concerned about one or more specific issues related to science. The attentive public for science would not be very large, but its

size might be expected to increase as the educational level of the populace continues to rise. Since
the attentive public generally utilizes specialized
information channels in addition to regular media,
a large proportion of the attentive public for organized science would probably read Science, Scientific American, and/or journals in various scientific
and professional areas.

For the most part, the mass public can be expected to hold few structured attitudes toward organized science. Following Converse (1970), the mass
public would display little science information when
open-ended questions are utilized. At the same
time, it should be noted that special topics can
generate short-term interest and the mass public may
become politically active on a given topic. The
bitter local fights over fluoridation that punctu-
ated the 1950s and early 1960s are an excellent
example (Mausner and Mausner 1955; Kirscht and
Knutson 1961; Davis 1960; Gamson 1966). There is
little evidence that these flashes of interest have
had any long-term impact and that any significant
number of persons move into the attentive public for
science as a result of these single-issue fights.
The siting of nuclear power plants has generated
similar local disputes in numerous communities
(Nelkin 1971; 1974; Richardson 1974; Carter 1976;
Doderlein 1976).

Recalling the foreign policy parallel, however,
it should be noted that the mass public retains a
final political veto power and could choose to use
it if sufficiently aroused. The growth of environmental concern during the last decade is a good il-
lustration, and it should be noted that one or more
major scientific disasters (a major radioactive or
chemical disaster) could at least temporarily gel
mass public opinion into an antiscience position.
The ultimate role of the mass public in ending both
the Korean and Vietnam wars should not be lost on
the scientific community.

In general, the stratified model for the public
understanding of science may be expected to consist
of a large mass public with little interest in or
understanding of organized science, a small but
growing attentive public, and small and highly specialized elite and decision-making groups. In form
and process, it is very much like the model for public understanding of foreign policy.

Expected Attitude and Behavior Patterns

With a stratified model of public attitudes toward
science and technology policy in mind, it is important to ask what attitudes and behaviors would distinguish the members of the attentive public for
organized science from the members of the mass public. On the basis of the literature reviewed in the
previous chapter and the experience of the authors
in working with the 1957 survey by the National
Association of Science Writers, four major sets of
expectations emerge.
     First, it is anticipated that the attentive
public for organized science will be more positive
in its view of the potential of science and technology to improve on life, currently and in the future.
This optimism may be based in part on a more extensive knowledge of the achievements of science and
technology during the last several decades, and it
may reflect a greater awareness of the scope and
extent of the present scientific and technological
enterprise. These positive views may also reflect a
generally more positive and optimistic view of life
and of society, or stated conversely, a significantly lower level of political and social cynicism.
     Second, it is anticipated that the members of
the attentive public for organized science will be
somewhat more aware of the risks inherent in many
scientific and technological activities and will be
more willing to accept risks than members of the
mass public. This is not to suggest a dichotomy in
which the attentives are all risk takers and the
mass public rejects all risks, but rather to suggest
that the proportion of persons willing to trade a
moderate level of risk for potential benefits will
be significantly larger in the attentive public than
in the mass public.
     Third, it is anticipated that the attentive
public will be more aware of the central role of
basic research in organized science and that the
members of the attentive public will be more supportive of federal expenditure for basic research
than the members of the mass public. This difference may reflect a higher level of knowledge about
the basic research/applied research interface on the
part of the attentives, and perhaps a greater willingness to make longer-term investments based on the
optimism for science and technology discussed above.

Finally, it is anticipated that the attentive
public will view the members of the scientific community as highly credible information sources about
science- and technology-related policy more often
than will members of the mass public. As the previous review suggested, there is a lingering wariness
about science and scientists among a significant minority of the American people, and it is reasonable
to expect those persons least interested in and
least informed about science issues -- the mass public -- to be the least trusting of scientists and
scientific organizations for information on policy
matters.

In summary, it is anticipated that the members
of the attentive public for organized science will
be significantly more positive toward science and
technology, more accepting of risk, more supportive
of basic research, and more trusting of the scientific community than the members of the mass public.
These are important differences and their potential
impact on the American political system is substantial.

II

The Data Base

4 The 1978 National Public Affairs Study

The data reported and analyzed in this monograph are based on the 1978 National Public Affairs Study (NPAS). The NPAS was designed to study attentiveness to a set of four issue domains -- science and technology, foreign policy, economic policy, and civil rights -- and to investigate the social factors that are important in determining attentiveness to these topics. The inclusion of multiple issue domains served the function of presenting the students with an array of topical issues so as not to suggest that they ought to be interested in any particular issue. This approach, it was hoped, would allow a more accurate assessment of the respondents' real interests and opinions by avoiding the arousal of evaluation apprehension that may occur when respondents are asked to indicate their interest in or knowledge of a single topic.

THE NPAS QUESTIONNAIRE

The NPAS questionnaire was designed to measure three aspects of issue attentiveness: (1) interest in a topic, (2) knowledge about the topic, and (3) regular acquisition of information regarding the topic area. (The instruments used in the NPAS are included in Appendix A.)

The students' interest in each topic was measured in two ways. First, each respondent was asked to read 32 headlines that might appear in a newspaper or magazine and to indicate which headlines he

or she would "definitelv read," "probably read,"
"probably not read," or "definitely not read."
Headlines reflecting each of the four issue areas
were included plus headlines dealing with sports and
fashion news. Second, at the beginning of each issue section of the questionnaire, the student was
asked whether he or she was "very interested,"
"somewhat interested," or "not very interested" in
the issue covered in that part of the questionnaire.

Knowledge of each of the four topic areas was
ascertained by asking the students to define a number of concepts that might be encountered in reading
about the issue. A count of the number of concepts
that were correctly defined by each student provided
a measure of the extent to which the student might
be able to comprehend information and participate in
informed discussion on each issue. In the area of
science and technology, the concept definition battery was supplemented with a set of items asking
each student to indicate what arguments he or she
had heard in favor of and against the construction
of additional nuclear power plants. The student's
ability to recall arguments on either side of this
particular technological controversy was taken as an
indicator of the extent to which the student was
attentive to current discussions of issues relevant
to science and technology.

Frequency of exposure to information regarding
public issues in general was measured by a series of
questions about the media each student used as his
or her "most important source of information about
current news events," the number of times each week
the student read a newspaper, the news magazines the
student read on a regular basis, and any specialized
news magazines the student might have read.

The remainder of the questionnaire dealt with
specific attitudes and opinions of respondents in
each of the four issue domains and several factors
suggested by the social science literature as potentially useful in understanding issue specialization.
These items included personality dispositions, religious beliefs, political orientations, family background, school courses and experiences, and educational and occupational aspirations.

POPULATION AND SAMPLE

The target population for the NPAS was students enrolled in 1978 in public high schools and undergraduates in public and private college in the continen-
tal United States. To deal with the noncomparabili-
ty of grades included in different high schools, the
high school population was limited to students in
the tenth, eleventh, and twelfth years. To deal
with the unforeseen difficulties of identifying the
population of "young adults" enrolled as full-time
students in community colleges, the NPAS population
was restricted to students in baccalaureate and
graduate degree-granting institutions. The high
school and college populations were sampled by separate, but comparable multistage cluster procedures.

The High School Sample

To ensure that high schools from various kinds of
communities were included in the sample, school districts in the United States were stratified into
three geographical types: central city districts
within Standard Metropolitan Areas (SMAs), suburban
districts within SMAs, and all non-SMA districts.
Table 4.1 displays the sampling design within these
strata.
    A target of 36 schools was set to provide suf-
ficient heterogeneity in the sample, and a target of
3,000 students was selected to allow multivariate
analyses of various developmental patterns. Accordingly, a total of 36 schools was selected across
strata based on the proportion of students within
each strata and the average size of the schools in
the strata. The final sample included eight high
schools from urban districts, 14 high schools from
suburban districts, and 14 high schools from rural
districts. The actual schools selected were determined by sampling school districts on a probability-
proportional-to-size of the schools within each
district.
    As a result of these procedures, four schools
from each of two urban districts were selected along
with 14 schools from eight suburban districts and 14
schools from seven rural districts.
    Each school drawn in the sample was contacted
and asked to participate in the study. All but

Table 4.1. Sampling Design

HIGH SCHOOL SAMPLE

Student Population by SMSA	Number of Districts Sampled	Number of Schools Sampled	Total Schools Responding	Number of Students Sampled	Number of Students Responding	Part. Rate
Urban	2	8	7	805	669	83.1
Suburban	8	14	14	1610	1101	68.4
Rural	7	14	13	1450	1147	99.1
Total	17	36	34	3865	2917	75.5

Note: Unexpected problems that prevented data collection occurred at one school the morning of the testing date. It was too late in the school year to select an alternate.

We treated the schools in a county school district and the city high school in that county as if they were part of the same school district.

Two substitutions were made in single high school districts. The replacements were similar schools in close proximity to the originals.

58

Table 4.1 (continued). Sampling Design

COLLEGE SAMPLE

	Number of Colleges Sampled	Number of Schools Responding	Number of Students Sampled	Number of Students Responding
4-year colleges/ universities with- out any doctoral programs	15	15	1016	533
4-year colleges/ universities with at least one doc- toral program	15	13	857	496
Total	30	28	1873	1029

Note: We dropped one college because only one student responded and a second was excluded because of a belated refusal to participate caused by local problems with the privacy act.

three schools agreed. Two schools indicated they
were already overcommitted as testing sites for
their respective state educational agency and could
not participate in another study. For each of these
schools, an adjacent district and school was substituted based on judgments made about the similarity
of the substituted district to the originally selected district. The third school, in a rural district, refused to participate in the study, and no
reasonable substitute could be found. This school
was dropped from the sample, and the rural strata
was underrepresented in the actual sample. The
weighting procedure described below was used to
restore appropriate balance to the final data set.
    The urban schools were also underrepresented by
one school. After agreeing to participate in the
study, unforeseen problems arose on the day of questionnaire administration at one school, making it
impossible for data to be collected. It was too
late in the school year to substitute another school
from the district. Consequently, one urban school
district is represented by three rather than four
schools, but this underrepresentation was also corrected by the weighting procedure.

                    The College Sample

A similar multistage cluster design was used to obtain a sample of college students. Colleges were
first split into two strata -- baccalaureate institutions that did not offer any doctoral programs and
doctoral-granting institutions. Next, 15 baccalaureate institutions and 15 doctoral institutions were
selected on a probability-proportional-to-size basis. Each institution was contacted to obtain directory information for either the entire student
body (if this information was available in published
form and differentiated undergraduates from graduate
students) or for a random sample of undergraduates.
The number of students selected from each institution was proportionate to the size of the undergraduate population within the strata.
    All 15 baccalaureate institutions that fell in
the original sample agreed to supply the directory
information; 13 doctoral institutions also agreed to
participate. The two refusals, one involving a
large student population, left the doctoral institu-

tions underrepresented in the final sample, but this
was also adjusted in the weighted data set.

DATA COLLECTION

Data were collected from late March to early June,
1978. The high school data were gathered by personal visits to each high school. Each questionnaire
was administered in a large-group session. Seventeen school districts and 34 high schools partici-
pated in the study (see Table 4.1). These high
schools were in 16 states from all regions of the
United States.
    The original sampling plan called for surveying
100 students from each high school. Officials in
some schools indicated that their daily absence
rates were as high as 15 percent. Thus, a random
sample of 115 students was drawn for each school,
and all 115 students were invited to participate.
Because participation in the study had to be strictly voluntary, the respondents were the "willing"
students who were in attendance the day the questionnaire was administered. All but two of the
schools were sufficiently large to have 115 students
in grades 10-12. A total of 3865 students was invited to participate, and 2917 of them completed a
questionnaire, producing a participation rate of
75.5 percent. When the total number of invitations
is adjusted for absentees (estimated at 15 percent),
the adjusted participation rate is 88.8 percent.
    Data were collected from the college students
by mailed questionnaires. Two follow-up mailings
were made approximately 15 and 30 days after the
original mailing. The first follow-up contained a
letter reminding the student of the importance of
his or her participation in the study and requesting
return of the questionnaire. The second follow-up
contained another reminder along with a second copy
of the questionnaire. A third follow-up was done by
telephone or mailgram to those who had not returned
their questionnaires by approximately two weeks following the second reminder. Half of the nonrespon-
dents received a mailgram, while half were contacted
by phone.
    A slightly higher proportion of students from
graduate institutions returned the questionnaire
(see Table 4.2). Overall, 55 percent of the students originally sampled returned usable question-

Table 4.2.   Response Analysis

	STRATA		TOTAL
	4-Year	Grad	
Original mailing	1016	857	1873
Bad addresses	37	26	63
Other disqualifications (e.g. graduated)	7	8	15
Corrected universe	972	823	1795
Usable questionnaires	533	496	1029
Partial questionnaires	9	0	9
Returned refusals	6	4	10
Nonreturn	424	324	748
Usable returns as percentage of corrected universe	55%	60%	57%

naires, and the response rate rises to 57 percent
when the original universe is corrected for persons
not reached or who were not undergraduates.

WEIGHTING THE SAMPLE

Responses to the National Public Affairs Study questionnaire were weighted to correct for dispropor-
tionate representation due to the sampling and response difficulties described above (see Table 4.3).
In the student population, 14.2 percent of high
school students attend high schools in urban dis-

Table 4.3. Comparison of Unweighted and Weighted Samples with Population

Strata	Population N	Population %	Unweighted Sample N	Unweighted Sample %	Weighted Sample N	Weighted Sample %
High School						
Urban	2,855,217	14.3	669	17.0	572	14.2
Suburban	5,062,304	25.3	1101	27.9	1018	25.2
Rural	4,988,175	24.9	1147	29.1	1007	25.0
College						
Nondoctoral institutions	3,103,924	15.5	533	13.5	625	15.5
Doctoral institutions	4,006,947	20.0	496	12.6	807	20.0
Total	20,016,567	100.0%	3946	100.0%	4029	99.9%

tricts. The sampling scheme and a participation
rate of 83.1 percent among urban high school students produced a slight overrepresentation of students from this strata. Consequently, each student
in the urban high school strata was weighted so that
the 669 urban high school students in the sample
represented 572 students in the weighted sample.
Comparable weightings in the other strata produced a
weighted sample in which each school is represented
in the weighted sample proportional to the size of
the school in the total student population. The
last column in Table 4.3 indicates the relative distribution of weighted responses across strata, which
in each case is within 0.1 percent of the proportion
represented by the strata within the total student
population. To obtain this degree of accuracy in
the weights, the total sample size had to be inflat-
ed slightly, from 3946 actual respondents to 4029
weighted responses.

CODING PROCEDURES AND REPLACEMENT
OF MISSING DATA

A small proportion of the questionnaires was returned with missing data in some of the items that were
central for the analysis. The following procedures
were used in an attempt to estimate what response
would have been given by the respondents to these
items:

1. Fifty-eight high school respondents did
    not indicate their year in school but
    did indicate their age. The modal grade
    for students of similar age was used as
    a replacement for year in school. High
    school students for whom no age was
    given were not given an estimated year
    in school, and no estimate was made to
    replace the missing value for college
    students who did not indicate their year
    in college.

2. Twenty-one college students completed
    and returned the questionnaire but indi-
    cated they had graduated in January 1978
    and were no longer students. These 21
    students were reclassified as seniors
    and kept in the analysis.

3. One hundred and five students did not
    indicate their sex but gave multiple re-
    sponses regarding interests and occupa-
    tional aspirations which could be used
    to unambiguously classify each as to
    sex. Two students for whom responses to
    other questions did not give a clear in-
    dication of their sex were removed from
    the analysis.

Missing data on other items that were used only
within multi-item indices in the analysis were re-
placed using the following procedures:

1. If the respondent answered a majority of
    items contributing to a single index,
    the average of these responses, rounded
    toward the modal response, was used as
    the replacement value for missing data
    within the index. As a result of this
    procedure, index scores are based on the
    average response to all items given by
    the respondent for those respondents who
    answered a majority of the items measur-
    ing any given construct. For most of
    the items, the total amount of estimated
    data is less than 2 percent.

2. If the respondent did not answer a ma-
    jority of items contributing to a single
    index but did answer most of the items
    in the same section of the question-
    naire, a nonresponse was taken as indi-
    cating that the respondent was "uncer-
    tain" or had "no opinion" regarding the
    missing items. In these cases, the
    "uncertain" or "no opinion" response was
    used as a replacement for the missing
    data. Less than 1 percent of the items
    were estimated in this manner.

INDEX CONSTRUCTION

Multi-item indices were constructed on a wide range
of variables used in the analyses of the NPAS data.
The rationale behind the use of indices rather than
single items was twofold. First, multi-item indices
tend to provide more reliable indicators of underly-

ing constructs than do responses to single items.
The internal consistency of the indices attests to
the items measuring common variation, as indicated
by the correlations among the item responses. Second, multi-item indices enhance variation across the
dimensions they measure. While a single item may
have only three possible responses -- say, agree,
disagree, and no opinion -- two similarly scored
items produce the possibility of nine distinct response patterns. Indices based on the combination
of two or more items, therefore, provide the possibility for greater variation that may be correlated
with and helpful in explaining responses to other
indices.

       To build multi-item indices, however, one must
use statistical procedures that are consistent both
with the construct one is trying to measure and with
the nature of the data one has to measure the construct. The procedure used in the present study
involves evaluation of the intercorrelations among
items thought to measure a given construct using the
ordinal correlation coefficient gamma. Gamma uses
only the ordinal information contained in item responses to measure the extent of intercorrelation
among a set of items. The size of the gamma
coefficient is interpretable as the proportional-
reduction-in-error (PRE) in guessing the respondents
order in pairs, on one item, given information about
their order on another item. The underlying model
against which gamma is a measure involves an ordinal, cumulative structure. Perfect prediction of
the order on one variable, given the order on the
second, produces a perfect correlation (1.0) indicating a 100 percent reduction-in-error. This coefficient of 1.0 is a reasonable indicator of the
extent of correlation if one is interested in asking
how closely two items approximate a cumulative response structure indicative of measurement of a
single dimension by the two items. Use of the gamma
coefficient, therefore, provides evidence regarding
the existence of an underlying dimension that is
common among the items without making any use of the
item responses as providing metric information regarding the measurement of that underlying dimension. Such a result appears to be consistent with
the present state of development of many of the
items used as indices of the attitudinal and personality constructs used in the present research.

       A second statistical technique, factor analysis
of the gamma matrices, was employed extensively in

building factor scales based on counts of responses to those items evidencing the greatest amount of variation in common with the underlying factors. Since the gamma coefficient is interpretable as a proportional-reduction-in-error statistic, where the proportion varies from 0 to 1.0, the factor loadings resulting from a factor analysis of the gamma matrix have analogous interpretations relative to the underlying factor. High loadings indicate items that are strongly correlated with other items in the matrix due to the commonality associated with a given factor, while lower loadings indicate weaker association among the items due to that factor. In general, oblique rotations of extracted factors to oblimin criteria were employed to identify the dimension that was maximally correlated with a given set of items (Harman 1967). The items that loaded relatively highly on each oblique factor were then employed together in a combined index, respecting the additional criterion that no item was used in more than one index. Use of the latter criterion avoids the likelihood of inflating estimates of association among constructs due to correlation of specific (and error) variation associated with individual items.

As a final step in index construction, the ordinal nature of the item responses was respected by avoiding the adding or averaging of responses (except in the replacement of missing data discussed above). Indices were built from counts rather than sums of responses. Typically, a respondent is classified as "high" on a given dimension if the majority of his or her responses to individual items measuring that dimension would lead to classification as "high" on the individual items. Hence, the index score is an attempt to summarize the modal responses to the individual items rather than an attempt to take their arithmetic average. The latter approach, of course, produces distortions due to extreme scores on individual items if the underlying metric on which the responses are based is not accurately measured by the item score.

Maturation and the Measurement of Attitudes,
Values, and Personality Traits

In any study of socialization or maturation that ex-
tends over a number of years or examines a number of
age cohorts, the problem of changes within conceptual structures must be taken into account. In the
present study, in which age cohorts from early teens
to late twenties are being compared, this problem
has been dealt with, when necessary, by defining
attitudes, values, and personality traits with reference to the older, more mature respondents and
using these definitions comparably when measuring
the attitudes, values, and personality traits of
younger members of the sample.
     An example regarding the measurement of a spe-
cific set of personality traits should clarify this
procedure. In chapter eleven, an argument is made
linking Rokeach's theory of the open and closed mind
(Rokeach 1960) to the development of interest and
knowledge of public issues. To measure those dimensions of closed-mindedness identified by Rokeach as
important, the NPAS questionnaire contained ten
items taken from a 66-item battery developed by
Rokeach. Previous studies (cf. Suchner 1977) have
indicated that, among college students, these ten
items consistently index three dimensions that are
interpretable as consistent with Rokeach's theory.
Factor analysis of these same items on the present
sample of college students replicates this finding.
Factor analysis of these items on the high school
population, however, evidences a much weaker set of
intercorrelations among these items, indicating less
internal consistency in responses, resulting in a
relatively unstructured factor pattern. Given these
results, we might choose to say that college students evidence the personality structures Rokeach
identifies as open- and closed-mindedness, while
high school students do not. An alternative interpretation, however, is that both cohorts respond to
the individual items in an "open" or "closed" manner, but the college students, having had a longer
time to develop their tendencies, are more consistent in their responses. Given this interpretation,
it is not so much that high school students are neither open- nor closed-minded as that they, as individuals, tend not to be consistently this way across
item responses. As a result, when we use the structure of responses given by the college students as

the basis for defining the dimensions of a mature
tendency toward either open- or closed- mindedness,
we are making the assumption that the high school
students' responses to the individual items are measuring the same tendencies that are simply less fully developed. On the indices combining these item
responses, then, the high school students will show
less variation.

III

Attentiveness to Organized Science

5 The Structure and Development of Interest in Organized Science

While attentiveness requires a high level of interest and knowledge as well as a regular pattern of information acquisition, interest is viewed as the major stimulus to activity in the other two components. A review of the general literature concerning the psychology of motivation and of attention in general suggests that a high or growing level of interest may be expected before major time and effort investments are made in regard to any topic or activity. The interplay between interest and the other components of attentiveness will be discussed again in chapter seven, where the three components will be combined and the overall algorithm examined.

THE CONCEPT OF ISSUE INTEREST

While the term interest has a broad and general meaning in our language, its use in this study is specific, and it is important to start with a clear understanding of the basic conceptualization. For our purposes, interest is characterized by (1) the active following of a topic and (2) a selection or sorting process for the narrowing of the full range of issues into more limited ranges.

The active versus passive nature of interest is an important aspect of this conceptualization. It is not hard to find numerous questions in the literature of survey research that ask respondents to characterize themselves as interested or not interested in various topics and then utilize those

single-item responses as measures of interest. This
approach would allow, if not encourage, a respondent
to express a wider range of interest than he or she
might be able to sustain in reality. Caution should
be exercised, then, in the use of single-item direct
inquiries about interest in a topic or subject area.
The second important facet of interest as it is
conceptualized for this study is the selection function of the judgment. Given the impossibility of
being intellectually or emotionally attached to more
than a small portion of the total range of public
issues at any point in time, it is important to view
interest as a selection process and to define a measurement that captures this aspect of the concept.

A Two-Dimensional Measurement Approach

Following the conceptualization of interest outlined
above, a two-dimensional measurement approach was
adopted. To assess an active interest in science
and technology issues, the respondents were asked to
classify themselves as very interested, somewhat
interested, or not very interested in science and
technology issues. The use of this trichotomy allowed the respondent to select a middle course, increasing the probability that a very interested or
totally uninterested response reflects a more definite feeling than would have been possible with a
simple dichotomy. As noted in chapter four, the
1978 NPAS instruments were designed to assess interest in a number of different issue areas, assuring
that the respondent would not see the instrument as
placing inordinate emphasis on any given subject and
thus signalling a situationally desirable response.
Using the direct interest inquiry and looking
at the data over the grade range of the 1978 NPAS,
it appears that the major difference is between
those high school students planning to attend college and those not planning to go to college. While
the absolute number of students reporting the highest level of interest in science and technology issues declines slightly during the high school years,
almost twice as many college-bound students report a
high level of interest than did their non-college-
bound cohorts (see Table 5.1).
The data indicate that college students are
slightly more interested in science and technology
issues than their college-bound high school counter-

Table 5.1. Interest in Science and Technology
Issues by Grade in School and Educational Plans

| | | Level of Interest | | |
Group	N	Not Very	Somewhat	Very	Total
HS-No Coll					
Year 10	466	29%	51%	20%	100%
Year 11	392	29	55	17	101
Year 12	359	27	55	18	100
HS-Coll Bnd					
Year 10	342	14	46	41	101
Year 11	361	15	46	39	100
Year 12	395	18	47	35	100
College					
Year 13	254	13	40	47	100
Year 14	319	10	44	46	100
Year 15	386	6	44	50	100
Year 16	462	7	46	47	100
Total	3736	17	48	35	100

Note: Throughout this analysis, data will be reported frequently by the three groups of the educational
aspiration typology: high school students without
college plans, high school students with college
plans, and students enrolled in college. The head-
ings HS-No Coll, HS-Coll Bnd, and College will be
used for the three groups respectively.

parts and that the level of interest is relatively
stable across the four years of the baccalaureate
experience. It is also interesting to note that the
proportion of college seniors reporting no interest
in science and technology issues is only half of the
proportion of college freshmen making the same report, suggesting that the college experience may
provide sufficient exposure to science and technolo-

gy issues so as to stimulate some movement into at
least the moderate level of interest. The relatively small proportion of college students reporting no
interest in science and technology issues means that
there is a larger pool of moderately interested persons in the population and the electorate who, under
certain circumstances, could be stimulated to even
higher levels of interest. A very high proportion
of persons expressing no interest in the issue area
would have suggested that it would be extremely difficult to generate any broad public interest in any
facet of the science and technology policy area.

In view of the caution suggested above concerning single-item inquiries about interest, it is important to find a second set of measures that will
tap the selection process and provide a check on the
original interest report. For this purpose, the
1978 NPAS included a set of 32 headlines similar to
those that might be found in a newspaper or a general news magazine (see Appendix A, page 306). The
respondents were instructed to read the full set of
headlines and to indicate for each item if they
would definitely, probably, probably not, or definitely not read the item, taking into consideration
the amount of time they generally had for reading
material of this type. The 32 headlines included
topics ranging from foreign policy to women's fashions and professional football. The series also
included a set of items concerning basic science
research, biomedical topics, space exploration,
ecological concerns, energy issues, and weapons
research. In general, the headline series generated
substantial variance, with respondents displaying a
high level of selectivity in their choices and with
few items being selected by a high proportion of the
respondents.

To better understand the structure of interest
as reflected in the headline responses, a gamma matrix was constructed for the 32 items in the series.
Following the procedures outlined in chapter four, a
factor analysis was performed on the matrix for the
college population in the 1978 NPAS (see Table 5.2).
Seven factors emerged, accounting for 66 percent of
the total variance in the matrix. Since some of the
factors were strongly associated with each other, an
oblique rotation was employed. It is useful to review the full set of factors as a structural over-
view of the thinking of the college respondents in
the 1978 NPAS.

The first factor included all five of the items
concerning foreign policy that were included in the
set of 32. The headlines that loaded on this factor
were, in order of the strength of loading:

European Common Market
Sets Foreign Trade Policy

U. N. Secretary-General Reviews
Conflict in Southern Africa

Third World & Nations Seek
New Trade Treaties

New Arab-Israeli Talks
Hint Middle East Peace

New Italian Elections May
Change NATO Alliance

It is clear that the five headlines express an interest in foreign affairs and that persons reporting
an intent to read several of these headlines might
be classified as interested in foreign policy
issues.

The second factor includes five items that reflect an interest in space exploration and weapons
research and this factor has been labeled technology
interest. The items loading on this factor, in the
order of the strength of loading, were:

Air Force to Test
New Weapons System

Defense Secretary Reveals
New Long-Range Missile

Soviets Launch New
Unmanned Space Station

Space Scientist Asserts
Need for Manned Flight

Professor Reports National
Study of UFO Sightings

It appears that the common thread running through
these items is an interest in technology, especially
hardware-oriented technologies. In retrospect, it
would have been desirable to have included a number

Table 5.2. Factor Structure for 32 Headline Items for College Students

VARIABLES (How likely to read)	Oblique Factor Patterns							
	1	2	3	4	5	6	7	h
Foreign trade policy	.87	.02	.03	-.02	-.06	.10	-.02	.73
Conflict in southern Africa	.74	-.04	.03	-.03	.06	-.18	-.03	.66
New trade treaties	.71	.03	.05	-.03	-.15	-.06	-.05	.66
Middle East peace	.66	-.13	.02	.03	-.00	-.05	-.01	.51
Italian elections	.66	-.05	-.11	.03	-.00	-.05	-.15	.56
New weapons system	.17	-.63	-.22	.02	-.10	-.19	-.08	.67
New long-range missile	.23	-.63	-.20	-.01	-.15	-.14	-.02	.68
Soviets launch space station	.15	-.62	-.03	-.03	-.04	.06	-.18	.58
Need for manned flight	-.08	-.56	-.13	.13	-.06	.01	-.32	.67
Study of UFO sightings	-.15	-.52	.34	.02	-.06	-.00	-.00	.39
Preview of summer fashions	.09	.12	.65	-.08	.13	.04	.10	.44
New birth control device	-.15	-.08	.64	-.12	.02	-.22	-.08	.60
Cancer therapy drug	.14	-.14	.59	.06	-.08	-.10	-.21	.48
Value of breakfast cereals	-.19	.05	.46	.02	-.11	-.11	-.12	.34
Impact of unemployment	.10	.25	.36	-.00	-.12	-.34	-.10	.46
Strength of NFL	-.01	-.01	.04	.91	.03	.02	-.01	.81
Baseball favorites	-.07	.03	-.02	.85	.05	.01	-.04	.72

Table 5.2 (continued). Factor Structure for 32 Headline Items for College Students

VARIABLES (How likely to read)	Oblique Factor Patterns							
	1	2	3	4	5	6	7	h
Energy crisis	-.11	-.05	-.08	.06	-.70	-.01	.19	.53
Chemical pollution	-.09	.04	-.04	.08	-.69	-.06	-.30	.67
Pollution in Great Lakes	-.06	.04	.05	-.09	-.65	.02	-.24	.56
Solar energy	-.01	-.12	-.00	-.05	-.59	.01	-.13	.46
New report on oil reserves	.24	-.27	.03	.05	-.55	.10	.02	.57
Policy to fight inflation	.21	.12	.08	.18	-.38	-.22	.08	.44
Freedom-of-speech case	.06	-.06	-.01	-.02	.06	-.77	-.09	.62
Police-search procedures	.00	-.25	-.03	.03	-.01	-.67	-.00	.49
Affirmative-action case	.25	.19	-.08	-.01	-.01	-.51	-.15	.48
Rules on sex discrimination	-.02	.07	.33	-.13	-.08	-.50	.06	.51
Create new jobs	.15	.14	.18	.18	-.28	-.41	.14	.55
Need for welfare reform	.31	.17	.09	-.06	-.21	-.37	.09	.51
Basic science research	.08	-.02	.06	.02	-.15	-.02	-.73	.70
Need for research funding	.14	-.04	-.04	.03	-.07	-.15	-.71	.69
Human cell modification research	.03	-.15	.29	-.10	-.06	.07	-.53	.54
Percent total variance	26.1	12.1	10.2	5.9	4.6	3.8	3.4	66.1
Percent common variance	43.5	18.7	15.5	8.6	5.8	4.3	3.7	

of less hardware-oriented technologies in the headline series, but it seems clear that this factor
does to a large extent reflect at least one major
facet of interest in technology. The first four
items loading on the factor have been included in an
Index of Technology Interest, which will be used
throughout this study. The UFO item was not included in the index because it loaded somewhat less
strongly than the others and because the factor
captured only 39 percent of the total variance in
the item.

The third factor included the following five
headlines, in the order of the strength of loading:

Paris and London Editors
Preview Summer Fashions

Medical Team Reports New
Birth Control Device

University Scientists Close
to Cancer Therapy Drug

Nutritionist Questions Value
of Most Breakfast Cereals

Sociologist Describes Impact
of Unemployment on Family

While the subject matter of the headlines is diverse, it appears that it may be summarized as reflecting women's concerns or interests. The high
loadings of the fashion item, the birth control
item, and the cancer item suggest a predominately
female interest grouping and an analysis of the
gender of the respondents reporting that they would
probably read the items on this factor indicated
that approximately 85 percent of the readers of
these five items were women.

Conversely, the fourth factor contains only two
headlines, both concerning sports. The very high
loadings indicate that a person who read one of the
headlines was very likely to read the other one, and
vice versa. An analysis of the gender of the persons who were likely to read both of these headlines
indicated that about three-quarters of the readers
were men. While this factor might be thought of as
a male factor, the growing level of sports interest
among women suggests that it would be more useful to
label the factor "sports interest" and take note of

the predominately male character of the present
readership. The headlines loading on the fourth
factor, in order of the strength of loading, were:

>   Cosell Reviews Strength
>   of NFL for Next Season
>
>   Baseball Writers Set
>   Favorites for New Season

The fifth and sixth factors both involve an array of domestic political issues, but they load on
different factors. The headlines loading on the
fifth factor, in order of the strength of loading,
were:

>   Oil Company Executive
>   Assesses Energy Crisis
>
>   Congress Moves to Curb
>   Chemical Polution
>
>   Scientific Group Reports
>   on Great Lakes Pollution
>
>   Senate Witnesses Debate
>   Future of Solar Energy
>
>   Geologists Release New
>   Report on Oil Reserves
>
>   Senator Asserts New Policy
>   Needed to Fight Inflation

The six items loading on the sixth factor were, in
order of the strength of loading:

>   Supreme Court Rules
>   in Freedom-of-Speech Case
>
>   Judge to Rule on Policy
>   Search Procedures
>
>   Suit Filed to Test
>   Affirmative Action Rules
>
>   State Court Rules on
>   Sex Discrimination

President to Ask Congress
to Create New Jobs

Experts Assess Need for
Basic Welfare Reform

It appears that the primary difference between the
two sets of issues is the personal involvement dimension. The issues on the fifth factor are problems that concern groups of people. Inflation or
pollution, for example, do not affect only one person or a small number of people, but larger groups
of persons or whole areas or regions. In contrast,
the items on the sixth factor involve actions that
directly affect individuals, one at a time. While
there are undoubtedly group ramifications from all
of the items on factor six, the primary impact of
sex discrimination, unemployment, or welfare payments
is on the individual. Much of the recent literature
concerning public attitudes toward domestic politi-
al issues has attempted to discern a liberal-
conservative trend and has not been able to find a
pattern of organization along those lines. The
division of the domestic issues reflected by these
headlines suggests another axis around which citizens may think about political issues, and this perspective will be explored later in the analysis.
    The seventh factor included the following
items, in the order of the strength of loading:

President Releases Report
on Basic Science Research

U. S. Scientists Assert
Need for Research Funding

Scientist Explains Research
in Human Cell Modification

The three items reflect an interest in science and
the support of basic research, and the three items
loading on the factor have been used to construct an
Index of Interest in Science that will be employed
in our definition of attentiveness to organized
science.
    Looking at the full set of factors, it appears
that the interest structure of the college respondents is ordered and interpretable. The high level
of substantive cohesion among the items loading on
each of the factors suggests that the responses re-

flected real preferences and contained little random
variation.

In contrast to the ordered structure of interest for the colleqe respondents, an analysis of the
interest structure for both the non-college-bound
and the college-bound high school respondents indicated substantially less structure and less interpretable patterns. In both groups, foreign policy
items, domestic policy items, science items, and
other items all load on the same factor. Only the
sports factor remains constant for all three cohort
groups, an interesting reflection of the deep roots
of sports in American culture.

A Two-Dimensional Measure

In view of the separate factors for science and
technology described above, it is necessary to
construct two measures of interest. Following the
logic of the discussion concerning the hazards of
single direct inquiries, a two-dimensional measure
was constructed that incorporates both the direct
inquiry and the indirect interest expression inherent in the headline battery.

While a single direct inquiry is not an adequate solo measure, it is a complementary measure to
the evidence provided by the indirect headline measures. For example, if respondents reported that
they were "very interested" in science and technology issues, it is reasonable to expect that they
should have indicated that they would have definitely or probably read at least one of the three
science-related items in the headline series. If
respondents reported a high level of interest in the
direct inquiry but no reading intentions, it would
appear that they may have overestimated their relative interest in science and technology issues.
Conversely, if respondents reported that they were
"somewhat" interested in science and technology
issues and then reported that they would have definitely or probably have read two of the three science headlines, it is reasonable to infer that they
underestimated their relative interest in science
and technology issues and should have been classified in the higher cateqory.

Using this approach, respondents were classified as being high on science interest if they responded that they were "very interested" on the

direct inquiry and that they would definitely or
probably read at least one of the three science
headlines. Respondents who labeled themselves
"somewhat interested" in science and technology issues but who indicated that they would have definitely or probably read two or more of the science
headline stories were classified as high on science
interest for the purposes of this analysis.

In a parallel manner, a respondent who reported
a high level of interest in science and technology
issues on the direct inquiry and who read at least
one of the four technology headlines was classified
as high on technology interest for the purposes of
this study. Similarly, a respondent who indicated
that he or she was "somewhat interested" in science
and technology issues and who would have read at
least two of the four technology headline stories
was classified as high on technology interest.

DEVELOPMENTAL PATTERNS

Using these two-dimensional measures, the data from
the 1978 NPAS were utilized to study the patterns of
interest development during the high school and college years. The discussions below will focus on the
total study sample, stratified only by cohort group.
In later chapters on the roots of attentiveness, additional analyses of the patterns of interest for
selected subgroups of the sample will be analyzed.

The data from the 1978 NPAS indicate that approximately 46 percent of the young adults studied
qualify as interested in science issues by the two-
dimensional measured described above. Slightly less
than 30 percent of high school students not planning
to attend college met the interest criterion, while
approximately 47 percent of their college-bound con-
temporaries qualified as interested in science issues. Almost 60 percent of college respondents met
the interest criterion, suggesting that the college
experience may enhance the level of interest developed during the high school years by the college-
bound cohort (see Fig. 5.1).

Interest in technology issues follows a similar
pattern, but at a higher level of interest than science at each grade level. The NPAS data indicate
that approximately 58 percent of the total young
adult sample qualify as interested in technology,
about 12 percent more than expressed an interest in

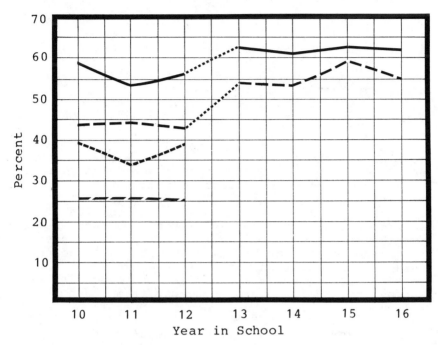

Interest in technology   HS-Coll Bnd & College ━━━━━
                          HS-No Coll   ----------

Interest in science   HS-Coll Bnd & College ━ ━ ━ ━
                          HS-No Coll   ━ ━ ━ ━

Fig. 5.1.   Percentage scoring high on science and
             technology interest.

science issues. Slightly over 40 percent of the
high school students not planning to attend college
met the technology-interest criterion, while just
over 60 percent of their college-bound colleagues
met this standard. As in the case of science interest, approximately 10 percent more of the college
respondents qualified as interested in technology
than did the college-bound high school cohort, suggesting a small but additive effect from the college
experience (see Fig. 5.1).
     In both the science and technology interest
distributions, it is apparent that the data fall
along three plateaus and that the patterns are stable within each major cohort. For example, while
college-bound high school students are more interested in both science issues and technology issues
than their non-college-bound counterparts, there are
no significant differences among any of the high

school grade levels within either group. Similarly, the year-by-year variation in the college cohort is minimal and not significant. This pattern suggests that the major difference in level of interest is captured by the difference in educational aspirations and that the college effect is both small and additive to the basic division in educational and career plans represented by the two high school cohorts.

As might be expected, interest in science issues and interest in technology issues are strongly related. An analysis of the interaction between the two measures of issue interest indicates that approximately 80 percent of those persons classified as interested in science issues are also classified as interested in technology issues, and that about two-thirds of those persons qualified as interested in technology issues are also interested in science issues -- a slight asymmetry (see Table 5.3).

Reflecting the platueau-like patterns discussed above, the modal group of students from the non-college-bound high school cohort expressed no interest in either science or technology issues, while the modal groups in both the college-bound and the college cohorts were interested in both science and technology issues. In all three cohorts, approximately 20 percent of the respondents report an interest in technology related issues, but no interest in science. Since the data are cross-sectional rather than longitudinal, it is not possible to determine if this is a stable interest group, or a preliminary stage of interest which is followed by the development of an interest in science issues.

LaPorte (1975b) has suggested that almost all persons experience technology in their everyday activities, from having an automobile repaired to using modern home appliances to dealing with computer-based information systems. In contrast, LaPorte argues that few people actually experience science or scientific research and that it is reasonable to expect most people to think about technology matters even when they use the word "science." The one major exposure that most people have to science -- other than those who choose it as a career -- is in the formal education system. When viewed in these terms, the data patterns observed above are more understandable. While all of the respondents in the study have experienced technology regularly, it is reasonable to expect that the non-college-bound student will receive the least

Table 5.3. Interest in Science and Technology Issues by Year in School and Educational Plans

Group	N	Interested in				
		Neither	Science only	Technology only	Science and Technology	Total
HS-No Coll						
Year 10	466	51%	6%	20%	24%	101%
Year 11	392	53	8	18	22	101
Year 12	359	50	6	20	24	100
HS-Coll Bnd						
Year 10	342	32	5	21	43	101
Year 11	361	34	7	18	41	100
Year 12	395	33	4	20	43	100
College						
Year 13	254	25	7	17	52	101
Year 14	319	24	9	18	49	100
Year 15	386	20	10	17	53	100
Year 16	462	23	10	18	50	101
Total	4029	35	7	19	39	100

exposure to formal science instruction. The
college-bound student may be expected to take more
science courses; many colleges require those courses
for college admission, and science items are included in many of the college admission tests. Further, the college-bound student will typically be
exposed to the most advanced science courses offered
in any given school. The college student who takes
the often-required introductory courses in one or
more science areas will receive a broader and more
sophisticated exposure to science. Needless to say,
the college student who majors in a science field,
or who takes a number of more advanced science
courses, will receive the most extensive exposure to
science. The proportion of students reporting an
interest in science issues closely reflects the
exposure patterns inherent in these three cohorts.

SPECIALIZED INTERESTS

In the preceding sections, the structure of interest
was defined according to the contours of the data as
reflected in the factor analysis of the headline series. This approach accepts the structure displayed
by the respondents and in this case, led to the division of the interest domain into a science component and a technology component. An alternative
approach would be to organize the headlines according to the preceptions of the analysts and to examine the patterns of interest within that structure.
In the NPAS, the headline series was constructed to
reflect six clusters of interest, more narrow in
definition than the two that emerged from the factor
analysis. While the factor-based interest definition will be used for the definition of attentiveness and in subsequent analyses in this monograph,
it is interesting and useful to examine briefly the
patterns of interest in the narrower domains of the
original design.
   The headline series was designed to incorporate
six separate facets of science and technology -- basic scientific research, space exploration, energy,
ecology, biomedical research, and weapons development. Two headlines were included for each area,
except energy, where three headlines were included.
Using the criterion of a definite or probable intent
to read at least one headline story in a cluster as
a measure of interest in that area, the 1978 NPAS

Table 5.4. Percentage Expressing Interest in Specialized Headline Clusters by Year in School and Educational Plans

Group	N	Headline Clusters					
		Basic Research	Space	Energy	Ecology	Bio-medical	Weapons
HS-No Coll							
Year 10	466	36%	50%	74%	63%	77%	59%
Year 11	392	37	48	66	59	80	55
Year 12	359	39	56	74	61	84	62
HS-Coll Bnd							
Year 10	342	47	69	80	68	88	67
Year 11	361	47	67	81	74	93	65
Year 12	395	50	73	81	71	89	73
College							
Year 13	254	55	78	89	79	94	72
Year 14	319	53	69	89	78	96	67
Year 15	386	62	74	90	80	93	64
Year 16	462	53	69	92	81	94	69
Total	3736	1775	2410	3039	2659	3303	2436

Note: Interest is defined as expressing a definite or probable intention to read at least one headline in the cluster.

data indicate that energy issues and biomedical research register the highest levels of interest at
all grade levels, regardless of educational plans
(see Table 5.4). Ecology issues rank third in the
level of interest, followed by space exploration,
weapons research, and basic scientific research, in
that order.

While the general patterns in the specialized
areas of interest are relatively stable and interpretable, it is useful to compare some of the items
in this structure with their location in the factor
structure. Two examples will suffice. First, the
two biomedical headlines included an item concerning
a research team announcing a new development in cancer research and an item about cell modification research, to reflect the contrast between basic and
applied research. The factor analysis indicates
that the college respondents did see the cell modification item as basic research, and it clustered
with the two basic research headlines. In contrast,
the cancer item loaded on the women's concerns factor, along with fashion, birth control, and breakfast cereals. It would appear that the respondents
viewed the item in terms of its personal impact
rather than as a reflection of a form of scientific
research.

Second, the headline series contained three energy items that were expected to either cluster as
a separate topic or to be viewed as one facet of a
more general science and technology factor. Instead, they loaded on a domestic issues factor that
also included inflation and the pollution items. It
would appear that these energy headlines were viewed
in terms of their domestic-political context rather
than for their scientific or technological import,
and the analysis will accept the order found in the
thinking of the respondents.

SUMMARY

In summary, the 1978 NPAS data indicate that a significant portion of the young adult population is
interested in either science issues, technology issues, or both. The major division in the study population appears to be related to educational and
career aspirations, with substantially higher proportions of college-bound high school students and
current college students qualifying as interested in

science or technology issues than the non-college-
bound high school cohort. Interestingly, the levels
of interest varied little within each of the three
cohorts by year of schooling.

6 The Levels and Sources of Science and Technology Information

The second and third major components of the attentiveness typology concern the level of current information and a pattern of regular information acquisition concerning organized science. Since the level of cognitive information and the sources and patterns of acquisition are closely related, this chapter will focus on both components. The first section of the chapter will outline the measures of science information and technology information used in the 1978 NPAS and examine the levels of information found. The second major section of the chapter will examine the sources of information regularly used by respondents, develop a typology of information acquisition concerning organized science, and examine the confidence of respondents in various information sources.

CURRENT INFORMATION ABOUT SCIENCE AND TECHNOLOGY ISSUES

The second major component of the definition of attentiveness to organized science outlined in chapter three is the level of current information. Given the separation of interest into science and technology components, the measures of current information have been designed to reflect the same division.

To measure science information, four substantive science questions were included in the 1978 NPAS, asking the respondent to define a molecule, an organic chemical, an amoeba, and DNA. The questions

were openended and the respondents were given two
lines to provide a simple definition. The responses
were coded as correct or incorrect. In general, the
coding conventions were lenient, but insisted on
certain minimal definitional components.
    The data display a surprisingly low level of
current substantive science information, especially
among the high school cohorts (see Table 6.1).
These data indicate that slightly over 70 percent of
the non-college-bound high school students could not
correctly identify any of the four items and that
none of the 1200 respondents in that group could
identify all four of the items. Since this group
will represent approximately half of the American
population in the next generation, the low level of
basic science information is particularly distress-
ing. The situation is not markedly better among the
college-bound high school cohort, with a full 40
percent of this group unable to correctly identify a
single substantive science item. Further, there is
no pattern of information growth during the high
school years in either cohort, suggesting a minimal
impact from formal science instruction during the
high school years.
    The college respondents performed much more
creditably on the substantive science items, with
slightly over 70 percent of the respondents correctly identifying two or more of the items. While the
growth of science information during the four college years, as implied by the four years of cross-
sectional data, is minimal, it should be recalled
that most college students take their introductory
science courses during the freshman or sophomore
years.
    In the calculation of attentiveness to science
issues, a respondent with two or more correct answers to this series of items is considered to have
a high score on science information. The application of this standard means that slightly less than
10 percent of the non-college-bound high school cohort qualify as informed about science information
while approximately 70 percent of the college student cohort meets the criterion.
    To measure information about technology issues,
the 1978 NPAS instrument included a pair of open-
ended questions asking the respondent to list the
positive and negative arguments that he or she had
heard concerning the construction of more nuclear
power plants. The survey was administered in the
spring of 1978, almost a full year before the Three

Table 6.1. Number of Science Knowledge Items
Correctly Identified by Year in School
and Educational Plans

Group	N	Number of Items Correct					Total
		0	1	2	3	4	
HS-No Coll							
Year 10	466	74%	20%	5%	1%	0%	100%
Year 11	392	68	23	7	3	0	101
Year 12	359	71	22	6	1	0	100
HS-Coll Bnd							
Year 10	342	40	24	2	12	1	99
Year 11	361	39	28	9	10	4	100
Year 12	395	41	25	1	11	3	101
College							
Year 13	254	12	16	0	24	19	101
Year 14	319	7	13	5	30	15	100
Year 15	386	10	14	2	33	21	100
Year 16	462	11	16	7	29	17	100
Total	3736	39	20	19	15	8	101

Mile Island accident, but following several nuclear
power referenda in various parts of the country the
previous year. Respondents in states in which a
referendum occurred may have had some advantage, but
the wide coverage of the issue by the national media
should have provided a reasonable level of information exposure to all of the respondents in the
study. The number of separate arguments provided by
a respondent were counted without judgment concerning the merit of the point or the extensiveness of
the argument description. A respondent could have
provided only positive arguments, only negative
arguments, or a mixture.
    The data indicate that a higher proportion of
respondents in all cohorts was aware of the issues
in nuclear power than was knowledgeable about basic

science concepts, and the absence of any developmental pattern during the high school years persists (see Table 6.2). The absolute levels of issues awareness, however, are disappointing at best. Almost 60 percent of the non-college-bound high school cohort were unable to name a single argument in favor of or opposed to the construction of additional nuclear power plants, as were approximately a third of the college-bound high school cohort. While it might be expected that this particular measure of technology issue information would be higher today in the aftermath of Three Mile Island and the dramatization of nuclear power accidents in motion pictures and television specials, it is likely that the level of issue awareness on other, less well-publicized technology issues has not increased markedly.

The data from the college cohort indicate that year in school is positively associated with the level of issue awareness. Generalizations from the cross-sectional data suggest that issue awareness grows during the college years at a statistically significant rate.

In the calculation of attentiveness to technology issues, a respondent with two or more argument citations is considered to have a high score on technology issue information. The application of this criterion means that approximately 25 percent of the non-college-bound high school cohort qualify as informed about technology issue information, while approximately three-quarters of the college student cohort meet this qualification.

In chapter five, it was observed that there was a substantial overlap in interest in science issues and interest in technology issues, with the same group of respondents constituting the majority of both groups. It is appropriate to raise the same question in regard to the information components of the attentiveness typology. Are the respondents who are informed about science also informed about technology issues, or is there some asymmetry? An analysis of the two information variables reveals a much stronger asymmetry between science information and technology issue information than was found in regard to science and technology issue interest. The data demonstrate the absence of either science or technology information among the non-college-bound high school cohort (see Table 6.3). Within the college-bound high school cohort, the level of both science and technology issues information is nega-

Table 6.2. Number of Nuclear Power Arguments
Cited by Year in School and
Educational Plans

Group	N	Number of Arguments Cited					Total
		0	1	2	3	4	
HS-No Coll							
Year 10	466	59%	15%	18%	4%	4%	100%
Year 11	392	60	14	16	6	4	100
Year 12	359	58	16	15	8	4	101
HS-Coll Bnd							
Year 10	342	31	16	33	4	7	101
Year 11	361	40	11	24	5	10	100
Year 12	395	33	15	26	7	9	100
College							
Year 13	254	21	3	34	0	22	100
Year 14	319	21	6	26	8	30	101
Year 15	386	15	4	21	5	35	100
Year 16	462	12	4	20	8	37	101
Total	3736	36	11	23	15	16	101

tively associated with year in school. Generalizations from these cross-sectional data suggest that
the level of information actually declines with
continued high school exposure for the college-bound
group!

For the college respondents, who demonstrated
much higher levels of both science and technology
issue information, the modal pattern was a high level of information about both science and technology.
Almost two-thirds of the college respondents were
informed about both science and technology. Only
about 10 percent of the college respondents were
uninformed on both measures.

Table 6.3. Percentage Scoring High on Science Knowledge and Issue Awareness by Year in School and Educational Plans

Group	N	Percentage Informed About				
		Neither	Science Only	Technology Only	Science and Technology	Total
HS-No Coll						
Year 10	466	71%	3%	23%	3%	100%
Year 11	392	70	5	21	4	100
Year 12	359	71	3	22	4	100
HS-Coll Bnd						
Year 10	342	36	11	29	24	100
Year 11	361	42	9	25	24	100
Year 12	395	42	6	24	28	100
College						
Year 13	254	14	10	14	62	100
Year 14	319	9	18	11	62	100
Year 15	386	9	10	16	65	100
Year 16	462	9	6	18	67	100
Total	3736	39	8	21	33	101

Note: To be classified as high, a respondent had to give 2+ correct science knowledge answers and cite 2+ nuclear power issues.

INFORMATION ACQUISITION

The third component of the attentiveness typology is a pattern of regular information acquisition about organized science. This component is important because it provides a measure of stability and continuity that is an important attribute of an attentive public. Without some measure of continuing interest and activity, it would be possible for a survey to find persons who have recently seen an article on a topic or heard a newscast and who would meet both the interest criterion and the current information criterion, but who may not be sufficiently interested to continue to follow the issue and maintain a high level of interest and information. While no single variable can completely capture the probability of continued interest and activity, the existence of a previous pattern of information acquisition on the topic or subject area is a reasonable surrogate measure. Given the cost of information acquisition in both time and resources, a pattern of regular information consumption about any specific topic or set of topics would appear to offer a reasonable probability of a continuing activity in regard to that topic or area.

Patterns of Media Consumption

The first step in the process of understanding the patterns of information acquisition is to examine the sources of current information reported by the respondents. In the 1978 NPAS, each respondent was asked about the frequency and focus of newspaper reading, the frequency and focus of magazine reading, and the frequency of television news viewing. Each respondent was also asked to indicate which of the media sources was his or her most important source of information. These items provide a very useful profile of the information consumption patterns of the young adults in the 1978 NPAS.

The data indicate that approximately 44 percent of the respondents usually read a newspaper at least five times a week and only 5 percent report the total absence of newspaper reading (see Table 6.4). The lowest levels of newspaper readership are found among the non-college-bound high school cohort and the highest rates of readership are reported by the

Table 6.4. Frequency of Newspaper Reading by Year
in School and Educational Plans

| | | Days of Newspaper Reading per Week | | | | | |
Group	N	None	1-2	3-4	5-6	Daily	Total
HS-No Coll							
Year 10	466	6%	37%	32%	11%	15%	101%
Year 11	392	6	38	24	11	21	100
Year 12	359	5	33	27	11	25	101
HS-Coll Bnd							
Year 10	342	4	28	23	17	28	100
Year 11	361	5	25	20	18	33	101
Year 12	395	2	23	23	19	33	100
College							
Year 13	254	3	26	27	18	26	100
Year 14	319	3	21	28	19	29	100
Year 15	386	4	25	19	16	36	100
Year 16	462	4	19	21	19	38	101
Total	3736	5	28	24	16	28	101

college respondents, reflecting the general profile
of interest and information levels discussed above.
Generalizing from the cross-sectional data, there
appears to be a gradual but steady growth in newspaper readership throughout the high school and
college years, with the highest rates of readership
being reported by college seniors.

While newspaper readership is suggestive of an
interest in public affairs topics, newspapers are
comprehensive publications that include a broad
range of items from gardening to foreign policy. It
is important to understand the focus of the newspaper readers in the 1978 NPAS. To allow an analysis
of the focus of newspaper readership, each respondent was asked to indicate if he or she usually read
each of several sections of the newspaper. An analysis of the results of this inquiry reveals a wide

range of interests. In general, those respondents
who read a newspaper five or more times per week
tend to read the local, state, and national news
sections of the newspaper. Less frequent readers
are much more likely to read local news than national news (see Table 6.5). As above, these patterns
are strongest among the college cohort and weakest
among the non-college-bound high school group.

It is interesting to note that the college respondents are slightly more likely to read the editorial section of the newspaper than the high school
cohorts, and that this tendency to read the editorial section appears to be positively associated with
year in school, suggesting a developmental pattern.
Since the editorial section usually focuses on policy debates in regard to public affairs issues,
this is another indication of targeted information-
seeking activity.

Working from the data discussed above, it is
possible to define a summary variable reflecting
newspaper readership targeted on public affairs issues. Respondents that reported reading a newspaper
five or more times per week and who indicated that
they usually read the national news section would
appear to be engaged in an important targeted
information-acquisition behavior. Accordingly,
persons meeting those criteria have been classified
as high on newspaper readership, and this division
of the population will be utilized later in the
construction of a composite information-acquisition
variable.

An examination of this summary measure of news
paper readership indicates that a gradual growth
pattern can be generalized from the cross-sectional
data (see Fig. 6.1). The patterns of newspaper consumption among the college-bound high school cohort
and the college cohort are almost linear and suggest
a continuing developmental pattern. In contrast,
the non-college-bound high school cohort reflects a
growth pattern during the high school years, but the
low starting point of this group leaves it far below
its college-bound counterparts. Further, since the
non-college group will not be receiving the stimula-
tion of the college experience, it is not clear what
type of pattern might be expected in the post-high
school years. It is an important question, however,
since this group will constitute approximately half
of the population of the United States in the next
generation.

Table 6.5. Sections of Newspaper Usually Read
by Frequency of Newspaper Reading by Level
of Schooling and Educational Plans

Days of Newspaper Reading per Week	N	Percent Usually Reading...					
		A	B	C	D	E	F
HS-No Coll							
1-2	411	50%	43%	77%	37%	36%	30%
3-4	328	61	35	82	51	47	33
5-6	130	66	39	90	54	57	37
Daily	222	65	30	83	55	59	32
HS-Coll Bnd							
1-2	264	51	34	85	47	50	28
3-4	237	60	34	91	58	63	32
5-6	190	69	37	87	65	68	28
Daily	339	73	29	86	66	74	33
College							
1-2	314	38	33	85	73	82	28
3-4	328	48	27	88	81	91	37
5-6	255	62	23	90	83	93	40
Daily	462	60	27	90	87	92	47
Total	3736	55	31	82	61	65	33

A = Sports Section   D = State News
B = Fashion Section   E = National News
C = Local News   F = Editorial Section

Note: The number of cases varied slightly by news-
       paper section, reflecting different rates of
       nonresponse. The number reported is the low-
       est number of any section in that row. The
       variation in nonresponse among the section
       categories averaged about 1 percent.

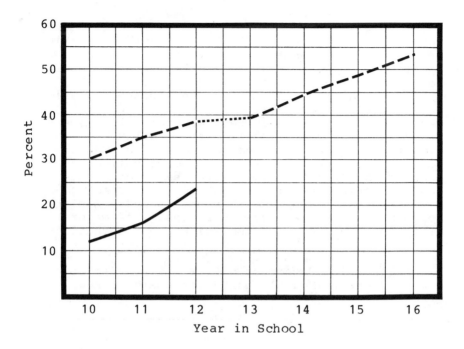

Fig. 6.1.   Percentage scoring high on newspaper
readership of national news.

        Magazines represent a second major media source
for public affairs information, including information about organized science issues. In the 1978
NPAS, each respondent was asked to report the names
of up to six magazines that he or she read regularly
and up to three additional magazines that were read
occasionally (see Appendix A, page 305). The name of
each magazine was coded and summary totals were produced by major classifications of magazines. For
the purposes of this analysis, it is useful to examine the readership of news magazines and science and
technology related magazines.
        The data from the 1978 NPAS reflect the same
general three-plateau form found in several of the
previous analyses. The lowest rate of news magazine
readership is reported by the non-college-bound high
school cohort, and the highest rate of news magazine

readership is found among the college cohort (see
Table 6.6). In contrast to the broader readership
of newspapers, approximately 60 percent of the total
sample and 50 percent of the college cohort report
not reading a news magazine regularly.
     The readership of science and technology ori-
ented magazines is substantially lower than the
readership of news magazines, with only 20 percent
of the total sample reporting the regular or occa-
sional reading of a science magazine (see Table
6.7). Since the 1978 NPAS instrument collected the
name of each periodical reported by respondents, it
was possible to examine the universe of reading
material consumed by the respondents and group it
appropriately for analysis purposes. The major
periodicals included in the science and technology

Table 6.6. Number of News Magazines Read by
Year in School and Educational Plans

Group	N	0	1	2 or more	Total
HS-No Coll					
Year 10	466	73%	19%	9%	101%
Year 11	392	76	18	7	101
Year 12	359	65	23	12	100
HS-Coll Bnd					
Year 10	342	64	25	11	100
Year 11	361	56	31	13	100
Year 12	395	58	28	15	101
College					
Year 13	254	49	36	15	100
Year 14	319	45	34	21	100
Year 15	386	50	38	12	100
Year 16	462	49	36	15	100
Total	3736	60	28	13	101

Table 6.7. Number of Science Magazines Read by
Year in School and Educational Plans

Group	N	Number			
		0	1	2 or more	Total
HS-No Coll					
Year 10	466	90%	8%	2%	100%
Year 11	392	88	9	3	100
Year 12	359	86	11	3	100
HS-Coll Bnd					
Year 10	342	72	19	8	99
Year 11	361	77	17	6	100
Year 12	395	80	18	3	101
College					
Year 13	254	80	13	7	100
Year 14	319	78	18	4	100
Year 15	386	73	19	7	100
Year 16	462	74	21	6	101
Total	3736	81	15	5	101

magazine category are Science, Scientific American,
National Geographic, Popular Science, Popular Mechanics, and Mechanix Illustrated. Magazines like
Car and Driver, photography magazines, and stereo
magazines were classified in the category of hobbies
rather than as science and technology.

The level of science magazine readership is
lower for the non-college-bound high school cohort
than the other two cohorts, but there is little
difference in science magazine readership between
the college-bound high school cohort and the college
cohort.

Television news represents a third major media
source of public affairs information. The importance of science news is increasingly recognized by
the major networks, all of which have strong science
news staff groups, and many local television sta-

tions have added a science editor to their news
teams.

   In the 1978 NPAS, each respondent was asked
about the frequency of television news viewing. The
data indicate that television news is viewed more
regularly by both of the high school cohorts than by
the college cohort (see Table 6.8). As will be suggested below, this may represent the greater accessibility of television for the high school student
who is usually living in the parental home than for
the college student who more typically would be
living in some type of institutionalized housing
facility. This interpretation is supported by the
observation that the frequency of television news
viewing by college students is positively associated
with year in school. This pattern would suggest

Table 6.8. Frequency of Viewing of Television
Newscasts by Year in School and Educational Plans

Group	N	None	1-2	3-4	5-6	Daily	Total
		Days of Television News Viewing per Week					
HS-No Coll							
Year 10	466	7%	22%	23%	16%	32%	100%
Year 11	392	5	28	22	13	32	100
Year 12	359	4	21	23	16	36	100
HS-Coll Bnd							
Year 10	342	3	15	27	23	32	100
Year 11	361	2	22	22	22	31	99
Year 12	395	3	16	28	18	35	100
College							
Year 13	254	9	29	26	15	20	99
Year 14	319	6	29	25	19	21	100
Year 15	386	8	23	29	13	28	101
Year 16	462	8	19	23	21	29	100
Total	3736	5	22	25	17	30	99

Table 6.9. Most Important Source of News
Information by Year in School and
Educational Plans

		Most Important Source					
Group	N	Radio	TV	News-paper	Maga-zine	Other	Total
HS-No Coll							
Year 10	466	34%	53%	11%	1%	1%	100%
Year 11	392	35	51	12	2	1	101
Year 12	359	32	53	13	1	0	99
HS-Coll Bnd							
Year 10	342	24	54	19	1	2	100
Year 11	361	29	52	17	0	1	99
Year 12	395	31	51	16	2	1	101
College							
Year 13	254	31	34	29	5	1	100
Year 14	319	37	31	28	4	0	100
Year 15	386	30	34	31	5	0	100
Year 16	462	25	41	29	4	1	100
Total	3736	31	46	20	2	1	100

that underclassmen, who are most likely to reside in dormitory type facilities, are least likely to watch the television news, and that upper classmen, who are more likely to live in apartments or other less institutionalized housing arrangements, are more likely to view the television news.

To examine the relative importance of each of the major media, the 1978 NPAS instrument asked each respondent to select the medium that represented his or her most important source of current news information. While the data show that television news is the modal information source for all cohorts and years, the overall pattern indicates a shift in media preference during the early college years (see Table 6.9). Slightly over half of both high school cohorts indicate that television is the most important news source, but this rate drops to just over

30 percent for college underclassmen. In a parallel
manner, college respondents report a much higher
rate of newspaper readership than either of the high
school cohorts. The rate of newspaper readership
remains stable during all four of the college years,
but the proportion of college upper classmen who report a primary reliance on television news is higher
than their underclass counterparts, suggesting the
accessibility explanation offered above. Since the
proportion of respondents reporting primary reliance
on newspapers and magazines does not differ significantly by year in school within the college cohort,
it would appear that the additional television view-
ers do not come from the print media, but rather
from radio and other news and information sources.
Needless to say, the absence of longitudinal data
makes these conclusions speculative.

Confidence in Information Sources

In the preceding discussion, the analysis focused on
the levels of media consumption and the respondent's
judgment about the relative importance of four major
media sources. To set these data in a broader context, it is important to understand which sources of
information the respondent judges to be the most
credible.
     In the 1978 NPAS instrument, each respondent
was asked to indicate how much he or she would
"trust each of the following people or sources to
give accurate and truthful information about science
policy issues?" The respondents were asked to cate-
gorize their level of trust as "a lot," "some," or
"not much." A "not sure" response column was also
provided to avoid forced choices not reflective of
real preferences (see Appendix A, page 311).
     An array of 13 persons and sources were provid-
ed, and the data indicate a high degree of selectivity by the respondents in all cohorts (see Table
6.10). The cross-sectional data also suggest a
number of important developmental patterns in the
assessment of information credibility. The highest
credibility rating in both high school cohorts went
to "A Congressional Committee on Science and Technology," but the level of confidence in the congressional committee was negatively related to the year
of school in the college cohort, suggesting a sharp
decline in confidence in the Congress during the

Table 6.10. Percentage of Respondents Expressing a High Level of Confidence in Selected Information Sources by Year in School and Educational Plans

Group	N	Percent Expressing "A Lot of Confidence in…"						
		Cong'l Cmte.	Univ. Prof.	E.P.A.	News Mag.	T.V. News	Chem. Co.Pres.	Radio News
HS–No Coll								
Year 10	466	54%	37%	36%	29%	36%	28%	18%
Year 11	392	60	34	37	29	31	32	17
Year 12	359	55	38	37	30	37	31	18
HS–Coll Bnd								
Year 10	342	65	50	52	32	30	31	15
Year 11	361	64	56	50	41	38	37	21
Year 12	395	57	49	49	40	36	31	18
College								
Year 13	254	42	50	47	47	32	14	17
Year 14	319	40	62	52	48	29	18	19
Year 15	386	32	58	50	38	23	12	17
Year 16	462	25	56	43	34	24	10	16
Total	3736	50	48	44	36	32	25	18

Table 6.10 (continued). Percentage of Respondents Expressing a
High Level of Confidence in Selected Information Sources
by Year in School and Educational Plans

		Percent Expressing "A Lot of Confidence in..."					
Group	N	Father	Teacher	Carter	U.N.	Mother	Student
HS-No Coll							
Year 10	466	19%	18%	22%	18%	20%	3%
Year 11	392	14	8	17	17	13	3
Year 12	359	16	14	18	14	15	2
HS-Coll Bnd							
Year 10	342	21	17	17	16	19	1
Year 11	361	16	21	15	17	12	2
Year 12	395	16	20	16	15	11	3
College							
Year 13	254	17	NA	12	10	8	4
Year 14	319	14	NA	12	8	9	3
Year 15	386	18	NA	11	10	7	4
Year 16	462	12	NA	9	11	6	5
Total	3736	17	17	15	14	13	3

college years. In contrast, President Carter rated
tenth among the 13 choices, and the proportion of
respondents expressing a high level of confidence in
the president also appeared to decline slightly
during the college years. This result is in sharp
contrast to the conventional wisdom in political
science that the president is the spokesperson for
the government and that this prominence emerges
early in the socialization process and remains
throughout the life cycle. In this case, given the
formal educational background of President Carter in
science and engineering, it is especially surprising
that only 15 percent of the total sample and 9
percent of college seniors reported a high level of
trust in the presidency to provide accurate and
truthful information about science policy.

The most plausible explanation would appear to
be that the prestige of the presidency has been
badly damaged by the Watergate scandal and other
adverse publicity. The 1978 NPAS asked an almost
identical question concerning confidence in the
sources to provide accurate and truthful information
about foreign policy, economic policy, and civil
rights issues, and in each of these other three
areas the appropriate congressional committee was
trusted by a higher proportion of respondents than
was President Carter. It is reasonable to suspect
that the prominent role of the Congress in the investigation of the Watergate events and the publicity focused on the impeachment hearings have
substantially raised the public esteem for the Congress and that these data reflect those perceptions.

The second highest level of confidence was
assigned to "a university professor." While only
about 36 percent of the non-college-bound high
school cohort expressed a high level of trust in a
university professor, a majority of respondents in
both the college-bound high school cohort and the
college cohort rated a professor as a trusted information source (see Table 6.10). College sophomores
were the most trusting of university professors.

The third most trusted source of science policy
information was the Environmental Protection Agency,
which was trusted by a majority of college-bound
high school students and a near majority of college
students. This is somewhat ironic in that the EPA
is an executive agency whose principal leadership is
appointed by the president, in whom the respondents
expressed considerably less confidence. Beyond the
relatively low level of understanding of American

government expressed in this contradiction, it would
seem that the presidency may be seen as a political
office and that the EPA may be seen as a technical
or scientific organization.

Approximately a third of the respondents placed
a high level of confidence in news magazines and
television news as information sources. The differences in trust between news magazines and television
news were not significant at the 0.05 level in either
of the high school cohorts, but college students
were significantly more trusting of news magazines
than the television news. Radio news was trusted by
approximately 18 percent of the sample, and this
level did not differ significantly by cohort or year
in school within cohort.

From these data, it would appear that the trust
displayed by the respondents is not in the media per
se, but rather in the persons and organizations
whose statements or actions are reported by the media, the congressional committee and the EPA being
two examples of this phenomenon.

Relatively few of the NPAS respondents placed
a high level of trust in information provided by
parents, other students, or the United Nations.
Students in the high school sample were asked to indicate their trust in information provided by their
teachers, and few respondents expressed a high level
of trust in their high school teachers for science
policy information. As noted above, President Carter ranked with this lower group in terms of the
proportion of respondents expressing a high level of
trust in him as an information source.

Looking at the proportions of respondents in
the total sample who express a lot of confidence in
the various news and information sources, one is
struck by the relatively low levels of confidence in
experts or opinion leaders in general. The most
trusted information source -- the congressional committee -- is trusted at a high level by only half of
the respondents, and that level of trust appears to
erode seriously among college students. The literature on public opinion has long asserted that persons who are relatively uninformed on a subject will
turn to more informed friends and colleagues who
will become informal opinion leaders. The Katz and
Lazarsfeld (1955) two-step flow-of-influence model
best epitomizes this perspective. The 1978 NPAS
data suggest, however, that there is little consensus about opinion leadership in the area of science
and technology policy.

This section began with an analysis of the frequency of various forms of media consumption and
then moved to an examination of the respondent's
views of the credibility of various information
sources, including the major media sources. Inherent in this discussion has been the assumption that
most respondents would hold a high degree of confidence in the medium that they rely upon as their
major source of news and current events information.
It is possible and desirable to test this assumption
empirically.

As discussed above, the 1978 NPAS asked each
respondent to evaluate 13 information sources in
terms of the degree of trust that the respondent
would have in the accuracy and truthfulness of information from that source. The instrument also
asked each respondent to indicate which medium was
the most important source of "information about current news events." While the evaluation question
involves sources other than the media, it did contain three of the four major media listed as primary
sources, and thus it is possible to examine the
levels of confidence in radio newscasts, television
newscasts, and news magazines according to the pri-
macy of that source to the respondent.

The data provide two conflicting impressions.
First, within a medium category (like radio newscasts), the highest proportion of persons expressing a high level of trust is usually the group of
respondents that relies upon that medium as its primary source. For example, 32 percent of the total
sample indicated that they had a high level of confidence in television newscast information, but 35
percent of those respondents who reported that
television is their primary source of current news
information expressed a high level of trust in television news information. Similar patterns appear
for radio newscasts and for news magazines (see
Table 6.11).

Second, in a contradictory pattern, the data
indicate that significant proportions of respondents
express high levels of trust in media sources other
than the one that they rely upon most heavily. For
example, 21 percent of the respondents reporting
that radio newscasts are their primary information
source express a high level of trust in radio newscasts, but 31 percent express a high level of confidence in television newscasts and 35 percent report
a lot of confidence in news magazines. While the
structure of the question was not mutually exclusive

Table 6.11. Confidence in Selected Information
Sources by the Most Important Information Source

| Most Important News Source | N | Percent Expressing "A Lot of Confidence in..." | | |
		Radio	Television	News Magazines
Radio	1224	21%	31%	35%
Television	1844	17	35	35
Newspapers	789	15	28	39
Magazines	95	21	24	39
Other	35	7	18	31
Total	3987	18	32	36

Note: Due to minor variations in response rates,
the actual N varied slightly for the three
confidence categories. For no category was
the variation more than 1 percent. The N
reported above is the lowest N for the three
tabulations.

(that is, a respondent could express "a lot" of
trust in none, one, two, or all of the sources), it
is not clear why only a fifth of the persons who use
radio news have a high level of confidence in it, or
why those persons with higher levels of trust in
television news or news magazines do not turn to
those alternative media. In the case of the television news, it might be possible to argue that some
persons do not have easy access to television sets,
as was done above in regard to college students, but
an analysis of the variables in Table 6.11 by level
of school and educational expectations displayed the
same pattern shown in the aggregate table. Further,
the problems of accessibility that might apply to
television would not apply to news magazines; thus
the continuing reliance on less trusted sources is a
behavior not explained by the NPAS data or by ordi-
nary logic. Less drastic, but equally perplexing results occur in regard to the respondent's reporting
primary reliance on newspapers and television.

A Measure of Information Acquisition

Having examined the patterns of information consumption and the level of trust in selected information sources, it is now possible to turn to the construction and analysis of a measure of information acquisition for eventual use in the determination of the membership of the attentive publics for science and technology.

If the function of the information acquisition variable in the determination of attentiveness is to be a measure of a stable and continuing interest in a given subject area, it would follow that the patterns of information acquisition should show some evidence of sufficient concern to justify the expenditure of a reasonable level of time and effort. Further, it is important that the quality of the information obtained be high and that its scope be appropriate to the issue cluster in question.

Applying these criteria to the major media sources measured in the 1978 NPAS, the best measures of a continuing interest in public affairs would appear to be newspaper and news magazine readership. In addition to a general interest in public affairs, it is appropriate to include measures of interest in science and technology issues specifically and science magazine readership would appear to be the ideal measure for this purpose. Television newscasts were excluded from the measure because the widespread access to televised news requires little respondent effort and therefore does not represent evidence of a sufficient interest in public affairs to expend scarce time and resources to become and remain informed and because the scope and quality of televised news is generally lower than the print media. Similarly, listenership to radio news requires little effort by the respondent and is limited in the depth of its coverage in many formats. Most of the deficiencies of television and radio news derive from commercial formats and would not apply to public radio or television stations or to selected all news and information commercial stations. On balance, however, it was the judgment of the authors that reliance on print information represented the best measure of a persistence of information acquiring behaviors in the respondents.

To operationalize the measure, three variables were used. In the discussion above of newspaper readership, a combined variable was described that

incorporated regular readership and some attention
to national news. This measure will be used in the
final typology of information acquisition.

Similarly, in the discussion of magazine readership, a variable was created to measure the number
of news magazines read regularly. This variable was
dichotomized into those respondents who read no news
magazines and those who read one or more news magazines, and this dichotomy is the second variable in
the determination of information acquisition.

Finally, a measure of science magazine readership was discussed above. This variable was dichotomized into those persons who read no science
magazines and those who read one or more science
magazines, and this variable will be used in the
final calculation of information acquisition.

At this point, the question becomes how much
information must a respondent consume to be considered a stable and continuing member of the attentive
public for either science or technology. It seems
clear that it would be an excessively high requirement to demand the regular use of all three types of
information sources, and it is not clear that reliance on a single source is adequate either. For
example, would it be reasonable to classify a person
who regularly read only National Geographic as attentive to science and technology issues? Or, a
person who regularly read only Time magazine? And,
given the range of newspapers from the New York
Times to small town dailies that stress local news,
it would appear risky to depend on newspaper readership alone. As a result, a respondent was required
to report the use of at least two of the three media
sources outlined above to be classified as high on
information consumption.

When this measure is applied to the 1978 NPAS
sample, 26 percent of the total number of respondents qualify as regular consumers of information
(see Fig. 6.2). Unlike the three plateau pattern
observed in several of the single medium measures
discussed above, the combined measure shows a pattern in which the proportion of respondents scoring
high on information acquisition is positively associated with both the level of schooling and the year
in school. Generalizing from the cross-sectional
data, it would appear that both the high school and
college years witness a growth in regular public
affairs information acquisition, with those high
school students planning to attend college starting
at a significantly higher point than their non-

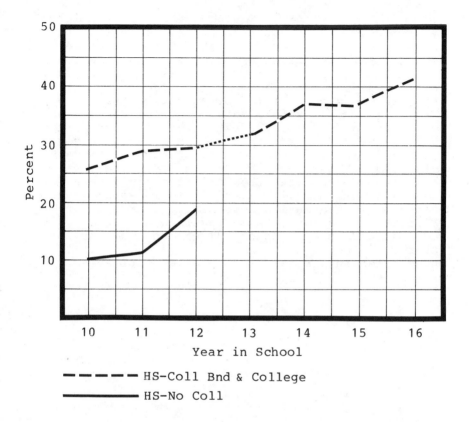

Fig. 6.2. Percentage of respondents scoring high on
science and technology information acquisition.

college-bound counterparts and maintaining that edge
throughout the high school years. Unlike some of
the disjunctive patterns noted earlier between the
college-bound cohort and the college cohort, the
pattern of information acquisition is almost linear
in nature.

SUMMARY

The measurement of attentiveness requires subsidiary
measures of interest, current information, and information acquistion patterns. This chapter has
examined the levels of current science information,
the levels of technology issue awareness, the
sources of current news information, the relative

importance of those sources, and the levels of trust
in selected information sources. From these analyses, two parallel measures of science information
and technology issue information were constructed
and analyzed, and these two variables will be utilized in the determination of attentiveness in
chapter seven. A single measure of information acquisition was developed, reflecting the less differentiated nature of the information sources in regard
to science and technology. While there are undoubtedly some publications that focus exclusively on
science or technology, they are relatively rare in
the media market at large and it would appear that
most of the information sources involving public
affairs information seek to cover both science issues and technology issues. The single measure of
information acquisition will be used in the following chapter in the calculation of attentiveness.

7 The Development of Attentiveness

The preceding two chapters have examined the structure of interest in science and technology issues and the levels and sources of information of these issues. Following Almond's model of an attentive public, it is the purpose of this chapter to review briefly the conceptual basis of the attentiveness typology, to examine the structure of attentiveness as displayed in the NPAS data, and to explore developmental patterns implied by the cross-sectional data available for analysis.

THE ATTENTIVENESS TYPOLOGY

In his work on public attitudes toward American foreign policy, Almond (1950) first proposed a stratified model of attitudes toward a low-saliency topic and subsequent work concerning public attitudes toward foreign policy has carried forward the attentive public concept, demonstrating its utility (Rosenau 1961, 1963, 1974; Hero 1959, 1960). This literature was discussed in greater detail in chapter three and that discussion will not be repeated at this point. It is important, however, to emphasize the conceptual foundation and structure of the three components in Almond's model.

In Almond's conception of the attentive public for foreign policy, interest and information are closely intertwined. If a person has little interest in a topic, it is unlikely that he or she will make an effort to acquire, analyze, and retain in-

formation about that topic. At the same time, when
issues are complex and remote from daily experience
and information is difficult to acquire, it is unlikely that an interest will arise on the part of an
individual or that a passing interest will be sustained. Almond describes the problem of public attentiveness to foreign policy:

> Under normal circumstances the American
> public has tended to be indifferent to ques-
> tions of foreign policy because of their
> remoteness from everyday interests and ac-
> tivities. When foreign policy questions
> assume the aspect of immediate threat to the
> normal conduct of affairs, they break into
> focus of attention and share the public
> consciousness with private and domestic
> concerns. It is not the foreign or domestic
> character of the issue which determines the
> accessibility of public attention, but the
> intimacy of the impact. From this point of
> view, foreign policy, save in moments of
> grave crisis, has to labor under a handicap;
> it has to shout loudly to be heard even a
> little....
>
> In addition to the pull of privatism, the
> very complexity of foreign affairs and the
> infinitesimal share of influence over the
> developments of world politics which the
> average American can realistically claim
> have the effect of discouraging interest and
> participation and providing justifications
> for indifference and lack of information
> (pp. 70-1).

While there is undoubtedly a certain amount of
circularity between interest and information for all
persons, it may be argued that the relationship between interest and information varies significantly
during the course of the life cycle. For the younger person who is still in high school or college or
even more advanced studies, exposure to information
on public affairs issues of all types is relatively
high, and it is more likely that information will
stimulate interest during this period than vice
versa. In contrast, for those adults outside the
formal educational system and whose employment does
not require substantial public affairs awareness or
involvement, the cost of information in time and

effort is high and exposure may be seen as highly
focused and purposive. In this phase, it may be
argued that interest becomes the dominant factor and
that the effort needed to acquire information is
made only when the level of interest is or becomes
relatively high.

In a similar manner, the levels of science and
technology issue information and the sources and
patterns of acquisition are intertwined. As Almond
(1950), Rosenau (1963, 1974), and others have demonstrated, the level of information can vary in the
short run, depending upon events of the day. For
example, if a survey were conducted just after a
nuclear accident like the Three Mile Island event,
it would not be surprising to find high levels of
information about nuclear issues. Conversely, if a
survey were conducted on a low-saliency topic during
a period devoid of special events, it is likely that
the level of information found would be much lower
for most respondents. The concept of an attentive
public requires the identification of respondents
who are as likely to be informed during noncrisis
times as during crises. The best approach would be
a longitudinal study to measure the levels of interest and information in the same respondents at various periods of time, but the resources available for
the 1978 NPAS mandated a single cross-sectional
study and it is that data base that is available for
analysis. In this context, the acquisition component may be seen as an effort to identify those respondents whose current information acquisition
habits appear to promise a continuing high level of
information.

The combination of the current information level and the patterns of information acquisition provide a good indicator of the information aspect of
attentiveness. If a respondent has a high score on
current information and a high score on information
acquisition, it is very likely that that individual
will continue to be informed in the future. For the
respondent with a high level of current information
but without an acceptable pattern of information
acquisition, there is no assurance of future information levels. The present high score on information could reflect information gained from a single
class or from a short period of high media exposure,
but there is no evidence of a pattern of behavior
that would assure new and timely information acquisition as conditions and events change. It is
possible that an individual with a high level of

current information will either develop the needed information acquisition habits or be able to keep current by other means, but this outcome is not as probable as for the respondent who is presently scored high on both information and acquisition.

The combination of a high score on information acquisition and a low level of current information is more troublesome. There are several plausible explanations. Since the variables used in the information-acquisition measure tended to reflect public affairs information consumption in general terms and only partially specific science or technology information consumption, it is possible that a person with a high information-acquisition score is busy acquiring public affairs information about issues other than science and technology. In this case, it is appropriate to exclude the respondent from the attentive public for science and technology. Alternately, it is possible that the range of information asked in the 1978 NPAS was too limited to capture the full range of science and technology interests and that some respondents are very well informed about other aspects of science and technology than those covered in the survey. To the extent that this occurs, the measure of attentiveness is insufficiently comprehensive. Several of these possibilities will be examined below.

In summary, the Almond model of attentiveness demands that a respondent rank high on interest and information and display an adequate pattern of information consumption about public affairs issues in general and the topic in question specifically. A number of respondents ranked high on one or two of the attentiveness components, but that result is not adequate for classification as a member of the attentive public for science or technology issues, or for attentive publics for other specialized interests. Those respondents scoring high on two of the indicators may be viewed as potential members of the attentive public and, at the close of this chapter, a number of conditions that might serve to stimulate the additional behaviors or attitudes needed for attentiveness will be discussed.

THE STRUCTURE OF ATTENTIVENESS

Having reviewed the conceptual basis of the attentiveness typology, it is useful to examine the structure of the three components of attentiveness as observed in the population in the 1978 NPAS. Following the division between science issues and technology issues discussed in the previous chapters, the analysis of the structure of attentiveness will focus first on attentiveness to science issues, then on attentiveness to technology, and finally propose a joint measure of attentiveness to either science or technology issues.

Following Almond's original model, attentiveness to science issues is defined by a high rank on science interest, a high rank on current science information, and a high score on information acquisition. The data from the 1978 NPAS indicate that only 11 percent of the total sample meet this criterion (see Table 7.1). An additional 14 percent of the respondents scored high on interest and information, but failed to meet the acquisition standard. Sixteen percent indicated a high level of interest, but failed to display either the requisite current information or a pattern of regular information consumption. It is likely that this additional 30 percent of the sample represent the pool for additional recruitment into the attentive public for science issues.

About 4 percent of the respondents scored high on both current information and acquisition, but failed to display an adequate level of interest. It is likely that persons meeting the acquisition requirement are interested in some set of public affairs issues, but the low score on interest suggests that the focus of their interest is not science. Another 10 percent scored high on information, but failed both the interest and acquisition standards, suggesting that they have acquired a certain amount of science information from courses and other sources, but that they have little interest in public affairs in general and little interest in science issues in particular. On balance, these groups appear to be less likely to become attentive by the Almond definition than the groups discussed above.

In a parallel fashion, attentiveness to technology issues is defined by a high rank on interest in technology issues, a high rank on technology issue information, and a high score on information acquisition. The information-acquisition variable used for measuring attentiveness to technology issues is the same variable used to define attentiveness to science issues above. The data from the

Table 7.1. The Structure of Attentiveness to Science
Issues by Year in School and Educational Plans

Group	N	Percent Scoring High on ...								
		I,K,A	I,K	I,A	K,A	I	K	A	None	Total
HS-No Coll										
Year 10	466	0%	2%	4%	0%	24%	4%	6%	61%	101%
Year 11	392	1	2	5	0	21	6	4	60	99
Year 12	358	1	3	7	1	18	2	9	59	100
HS-Coll Bnd										
Year 10	342	8	12	8	4	20	11	6	32	101
Year 11	361	9	9	7	3	22	11	9	29	99
Year 12	394	9	13	10	3	15	9	9	32	100
College										
Year 13	255	18	26	6	6	9	23	2	11	101
Year 14	319	23	30	2	7	3	20	5	10	100
Year 15	386	23	30	3	7	8	16	4	10	101
Year 16	462	24	25	3	10	7	13	4	13	99
Total	4029	11	14	5	4	16	10	6	34	100

Note: I = interest, K = information, and A = acquisition.

1978 NPAS indicate that 15 percent of the sample
scored high on all three of the components and were
classified as members of the attentive public for
technology issues (see Table 7.2). About 22 percent
of the sample scored high on technology interest and
current information, but did not meet the requirements for a high ranking on acquisition. Another 16
percent of the respondents registered a high level
of interest in technology issues, but did not display a high level of either information or acquisition behaviors. As was suggested in regard to
science issues, these respondents with a high level
of interest and some information may be the best
candidates for growth in the size of the attentive
public for technology issues. While the overall
level of both interest and information was slightly
higher for technology than science, the general
pattern of the responses on the three components of
attentiveness are remarkably similar.
     In the previous analyses of interest in chapter
five and information in chapter six, it was observed
that there was a number of respondents who scored
high on both science and technology measures and
analysis revealed a substantial overlap in both
interest and information. In the same vein, it is
important to examine the patterns of membership in
the attentive public for science issues and the
attentive public for technology issues. The data
from the 1978 NPAS indicate that about 9 percent of
the total sample were attentive to both science and
technology issues, in contrast to 6 percent who were
attentive to technology issues only and 2 percent
who were attentive to science issues only (see Table
7.3). These data mean that almost all of the persons attentive to science issues are also attentive
to technology issues, and that a majority of those
attentive to technology issues are also attentive to
science issues, but there is a slight asymmetry in
the relationship.
     In view of the substantial overlap between at-
tentiveness to science issues and attentiveness to
technology issues, it is appropriate to consider a
joint measure to facilitate the analysis of those
respondents attentive to either topic. As was noted
in the preceding chapters, the substance and measures of science and technology are completely independent and there is often a great deal of confusion
in the minds of respondents about the difference between the two terms. And given the relatively small
size of the two attentive publics, it will often be

Table 7.2. The Structure of Attentiveness to Technology
Issues by Year in School and Educational Plans

Group	N	Percent Scoring High on ...								
		I,K,A	I,K	I,A	K,A	I	K	A	None	Total
HS-No Coll										
Year 10	467	3%	11%	3%	1%	27%	11%	3%	41%	100%
Year 11	392	4	13	3	1	20	8	3	49	101
Year 12	358	7	11	6	1	20	7	4	44	100
HS-Coll Bnd										
Year 10	341	17	25	5	1	17	11	2	22	100
Year 11	361	16	18	6	4	19	11	2	24	100
Year 12	395	14	24	8	6	17	9	3	20	101
College										
Year 13	253	21	36	5	4	7	15	2	10	100
Year 14	319	23	30	3	7	10	13	4	9	99
Year 15	385	27	31	3	6	8	16	1	8	100
Year 16	461	27	33	4	10	4	15	1	6	100
Total	4029	15	22	5	4	16	11	2	25	100

Table 7.3. Attentiveness to Science and Technology by Year in School and Educational Plans

		Attentive to ...				
Group	N	Science and Technology	Science Only	Technology Only	Neither	Total
HS-No Coll						
Year 10	466	0%	0%	3%	97%	100%
Year 11	392	1	0	3	96	100
Year 12	359	1	0	6	93	100
HS-Coll Bnd						
Year 10	342	6	2	10	81	99
Year 11	361	9	1	8	83	101
Year 12	395	8	1	6	85	100
College						
Year 13	254	16	2	5	77	100
Year 14	319	18	5	5	72	100
Year 15	386	19	4	8	69	100
Year 16	462	18	6	8	67	99
Total	4029	9	2	6	84	101

analytically useful to have a single attentive group
that will approximate an attentive public for organized science, as it was defined in chapter one.
      For these reasons, a combined measure was con-
structed and labeled the attentive public for organized science. For the total sample, approximately
17 percent of the respondents are classified in this
group. In Table 7.3, the attentive public for organized science is composed of the respondents in the
first three columns, or all respondents except those
in the "neither" column.
      In summary, the 1978 NPAS data allow the iden-
tification of an attentive public for science issues, an attentive public for technology issues, and
a more general attentive public for organized science. The size of the three groups is relatively
small, indicating the low saliency of science and
technology issues to most young Americans.

                THE DEVELOPMENT OF ATTENTIVENESS

      Although the 1978 NPAS data set is a single
cross-sectional set of measures, it is possible to
utilize the data as indicators of likely patterns of
development. The authors are fully aware of the
hazards of making definitive judgments about developmental patterns from cross-sectional data and wish
to emphasize the exploratory nature of the following
discussion. In view of the importance of the issues
involved in present and future science and technology disputes, it is imperative that well-designed
longitudinal studies of the developmental patterns
hypothesized below be undertaken at an early date.
      Looking first at attentiveness to science is-
sues, the data suggest that the presence or absence
of college aspirations marks a major division among
high school students, but neither group displays any
developmental pattern (see Fig. 7.1). For high
school students not planning to attend college, attentiveness to science issues is virtually nonexistent. A full 60 percent of the non-college-bound
students failed to score high on a single component
of the attentiveness typology. This would suggest
that this group, which will constitute half of the
electorate in future years, is not a fertile ground
for stimulating attentiveness to science and technology issues.

In general, the proportion of high school students planning to go to college who are attentive to
science issues is significantly higher than their
non-college-bound colleagues, but devoid of any developmental pattern (see Fig. 7.1). About 9 percent
of this group met the definition for attentiveness
to science issues, and another 12 percent met the
interest and information criteria, but failed on
the acquisition requirement. Only a third of the
college-bound group failed to score high on any of
the three components of attentiveness.
  In contrast to the two high school cohorts, the
college respondents in the 1978 NPAS were substantially more likely to be attentive to science issues
than the two high school cohorts and the pattern of
attentiveness suggests a strong developmental pat-

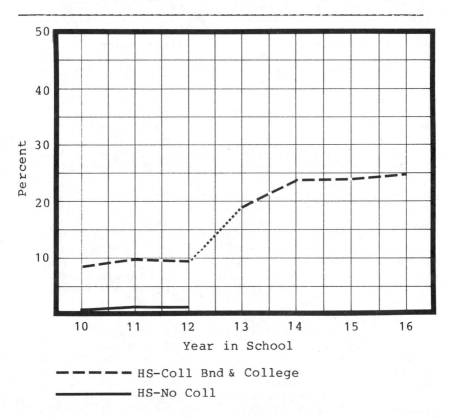

Fig. 7.1. Percentage attentive to science issues.

tern during the college years. The proportion of
college students attentive to science issues rises
from 18 percent for freshmen to 24 percent for seniors. For all four college years, another 25 percent of the respondents displayed a high level of
interest in science issues and a high level of current knowledge about those issues, but failed to
meet the acquisition criterion. It is possible that
some portion of these potential attentives manage to
keep informed through course-related readings and
other media not included in the acquisition definition used for this study and that they will become
attentive in the future as they move into normal
work and community activities. If this pattern were
to develop, as many as half of the college graduates
entering the electorate could be either actual or
potential members of the attentive public for science issues. This outcome stands in sharp contrast
to the almost total absence of interest among the
non-college-educated half of the electorate.

Turning to attentiveness to technology issues,
the 1978 NPAS data reflect the same three cohort
pattern that was observed in regard to attentiveness
to science issues, except that the absolute levels
of attentiveness are slightly higher (see Fig. 7.2).
The proportion of students attentive to technology
issues increases slightly among non-college-bound
students, rising from 3 percent for tenth graders to
7 percent for high school seniors. This suggests a
weak developmental pattern. About 11 percent of the
non-college-bound group displayed high interest in
technology issues and a high rank on current information, but did not engage in regular information
acquisition behaviors. Compared to the levels of
attentiveness and interest in science issues, attentiveness to technology issues is substantially higher for the non-college-bound group. This pattern is
very consistent with LaPorte's observations about
exposure to technology.

Approximately 16 percent of the college-bound
high school cohort qualified as attentive to technology issues and at least 20 percent more met all
of the criteria except acquisition (see Fig. 7.2).
The data indicate no developmental patterns within
this cohort for attentiveness to technology issues.

As with attentiveness to science issues, the
proportion of college respondents meeting the attentiveness requirements for technology issues is substantially higher than the college-bound high school
group and suggests a pattern of gradual growth dur-

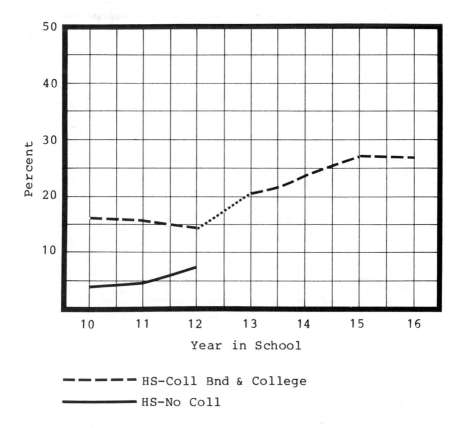

Fig. 7.2. Percentage attentive to technology issues.

ing the college years. The proportion of respondents classified as attentive to technology issues
rises from 21 percent for college freshmen to 27
percent for college seniors. Throughout the four
college years, an additional 30 percent of the respondents displayed high levels of interest and current information, but not the necessary patterns of
information acquisition. In a parallel manner, the
proportion of college respondents failing to score
high on any of the three components of the attentiveness typology declined from 10 percent for
freshmen to 6 percent for seniors.
     The 1978 NPAS data reveal several interesting
developmental patterns within the interaction of attentiveness to science issues with attentiveness to
technology issues (see Fig. 7.3). First, for the
non-college-bound high school cohort, the only ac-

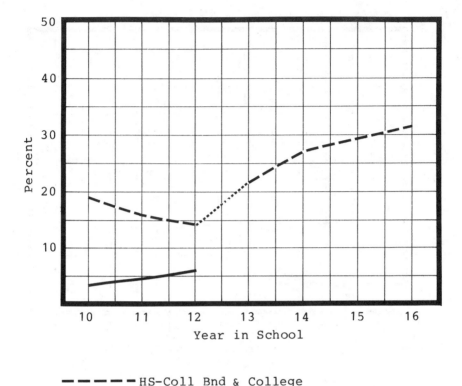

Fig. 7.3.   Percentage attentive to organized science.

tivity is a slight growth in the level of attentiveness to technology issues. Second, for the college-
bound high school cohort, the data suggest that
attentiveness to technology issues declines slightly
during the high school years and that the joint interest in science and technology issues grows corre-
spondingly (see Table 7.3). These patterns coexist
with a decline in attentiveness in general for the
college-bound cohort during high school. Following
LaPorte's observations about experiencing technology
normally and experiencing science only in formal
settings, these patterns would suggest that the
continuing and increasing exposure to high school
science courses may have the effect of stimulating
an interest in science issues among those already
interested in technology issues.

In contrast, in the college cohort, it appears that the proportion of respondents interested in both science and technology issues is relatively stable during the four years of college, but that there is some slight growth in the proportion of college students attentive only to science issues and in the proportion of students attentive only to technology issues (see Table 7.3). Without longitudinal data, it is not possible to track the entry of new attentives or the stability of previous attentives, but the cross-sectional data indicate that the proportion of persons attentive to organized science is very stable and that the proportion of one-issue specialists is growing, suggesting that the new recruits may be attracted to a narrower target of interest. Among college seniors, a full third of the respondents are presently attentive to science issues, technology issues, or both.

In summary, the 1978 NPAS data reveal substantial differences among the three educational cohorts utilized in the study, but relatively little in year-to-year developmental patterns. The data suggest a very strong college effect in the development of attentiveness, with a major jump during the freshman year and a pattern of gradual growth in proportion of attentives in the following years.

8 The Impact of Attentiveness on Science and Technology Policy Views

At this point, it is important to ask what difference attentiveness makes in an individual's general orientation toward science and technology and specific policy views. If attentives and nonattentives were to see most science issues the same, then there would be no clear advantage for identifying an attentive public. This chapter will examine general orientations toward science and technology issues, look at some specific policy questions, and explore the federal spending priorities of the attentive and nonattentive publics.

On the basis of the preceding discussion, what attitude differences should be expected between the attentive public and the nonattentive public? First, since the attentive public is by definition more interested and better informed, it is reasonable to expect that they ought to have clearer ideas about the benefits and risks of science and technology. On individual items, the attentives should be significantly less likely to indicate uncertainty or a "don't know" response.

Second, it is probable that the attentive public will be more positive toward science and technology and more capable of balancing risks and benefits. In general, it is difficult for an attentive public to be formed and to sustain itself over any period of time on a strictly negative basis. Most of the traditional interest groups in the United States are organized for the positive acquisition of benefits for their own clientele, and not primarily for the purpose of stopping or re-

straining some other group or activity. Given the
level of effort required to become informed about
science and technology issues, it is likely that
persons making that type of investment have a
generally positive effect for the scientific and
technological enterprise.
    Finally, if the attitudes of the attentives are
firmly rooted in an information network and an interest structure, it is reasonable to expect that
the attiitudes of the attentives will be more inter-
nally consistent and that their attitudes will be
more stable over time. Unfortunately, the absence
of longitudinal data prohibits good measures of stability, but it will be possible to examine the consistency of attitudes for the attentive and
nonattentive publics.

           GENERAL ATTIUDES TOWARD SCIENCE AND TECHNOLOGY

    It is appropriate to begin with an examination
of general attitudes held about science and technology issues in broad terms. As Hennessy (1972) and
others (McClosky 1964; Converse 1964) have argued,
most citizens have only fragile opinions on low-
salience topics and few attitudes. It is reasonable to expect that some respondents may hold
broad general attitudes toward science and technology, yet not be able to translate these broader
views into concrete policy choices in more specific
issues. For these respondents, the general attitudes become a receiving structure for new information and new issues and may, following the logic of
Allport (1954), become a cognitive screen for incon-
gruent information about science and technology.
    Following the principle of multiple measures of
attitudes whenever possible, the 1978 NPAS included
eight items that were designed to capture broad general feeling about the potential benefits and the
potential risks of science and technology. To better understand the structure of the eight items, a
gamma matrix was formed and a factor analysis was
performed. Two clear factors emerged, explaining
approximately 55 percent of the total variance in
the eight items. The two factors were correlated
at the level of -.37; thus the oblique rotation was
used for purposes of analysis and is reported in
Table 8.1.
    The two factors can be labeled risk and bene-

Table 8.1.   The Structure of Evaluative Attitudes
Toward Organized Science

Variables	Factor Patterns		h
	1	2	
1. Science makes our lives change too fast	.73	-.00	.53
2. Science means that a few people can control our lives	.62	.06	.36
3. Science breaks down ideas of right and wrong	.62	-.01	.39
4. We depend too much on science and not enough on faith	.55	-.10	.35
5. Science is making life healthier and easier	-.19	.74	.68
6. Scientific invention is responsible for our standard of living	.16	.55	.21
7. The fed. gov't. should spend more on science research	-.21	.48	.35
8. Science and technology have caused more good than harm	-.27	.38	.29
Percent Total Variance	38.1	16.6	54.7
Percent Common Variance	77.2	22.8	

fits, respectively. The four items loading on the
first factor all reflect some type of reservation
about science and technology and some perception of
a risk from those activities. The four items loading on this factor will be combined in a Risks of
Science and Technology Index. The four items loading on the second factor all deal with the promise
or the benefits of science and technology, and
these four items will be combined to form a Benefits
of Science and Technology Index.

As noted in chapter one, the concepts of benefit and risk are deeply imbedded in the elite and
attentive discussions of science and technology.
Given the decades of public confidence in high benefits and low or nonexistent risks, it is important
to begin this analysis with an examination of the
perceptions of benefits and risks held by the NPAS
sample of young Americans.

Attitudes Toward the Benefits of Science and Technology

The four items loading on the benefits factors
all reflect a positive orientation toward the scientific and technological enterprise in general,
but each item also incorporates slightly different
perspectives. It is useful, therefore, to start
with an analysis of the four items individually and
then examine the scores on a composite index.

The strongest loading item of the four was the
statement that science is "making our lives healthier, easier, and more comfortable." Approximately
three out of four NPAS respondents reported agreement with the statement, and a significantly higher
proportion of the attentives in all three cohorts
agreed with that statement than nonattentives (see
Table 8.2), but the differences among the three cohorts were not extreme. As was suggested in the introductory discussion, the attentives held somewhat
stronger views than nonattentives (i.e., a larger
proportion of attentives in every cohort was willing
to report strong agreement than their nonattentive
counterparts), and the proportion of respondents
giving an "uncertain" or "not sure" response was
higher for nonattentives than attentives in all
cohorts.

The second strongest loading item concerned the
impact of scientific invention on the American stan-

Table 8.2. Distribution of Responses to Item:
"Science is Making Our Lives Healthier,
Easier, and More Comfortable."

Group	N	SD	D	U	A	SA	Total
				Responses			
HS-No Coll							
Attentive	56	0%*	7%*	15%	55%	23%*	100%
Nonattentive	1161	5	17	20	45	12	99
HS-Coll Bnd							
Attentive	185	3	14	5*	50	28*	100
Nonattentive	913	4	14	13	50	19	100
College							
Attentive	417	1*	9*	11*	58	20*	99
Nonattentive	1005	3	12	16	55	14	100
Total							
Attentive	670	1*	11*	10*	56*	22*	100
Nonattentive	359	4	15	17	50	14	100

Note: In this and subsequent tables, * denotes a
difference significant at the 0.05 level.

dard of living, and this statement produced the highest levels of agreement of any of the four items
loading on the benefits factor. Within every cohort,
the proportion of respondents reporting a strong
agreement was significantly higher for attentives
than nonattentives (see Table 8.3). The highest level of disagreement was registered by respondents in
the non-college-bound high school cohort, a larger
proportion of whom may experience a less prosperous
standard of living than their college-bound counterparts and therefore may be less enthusiastic in response to statements about the high standard of
living in the United States.
     The third strongest item on the benefits factor
was a statement asserting that the federal government "should spend more money on science research."
The item was not designed to tap a specific and informed budgetary judgment, but rather a general feel-

Table 8.3. Distribution of Responses to Item:
"Scientific Invention is Largely Responsible for
Our Standard of Living in the United States."

| Group | N | Responses | | | | | Total |
		SD	D	U	A	SA	
HS-No Coll							
Attentive	56	2%	2%*	6%	35%	56%*	101%
Nonattentive	1161	2	6	11	45	36	100
HS-Coll Bnd							
Attentive	185	0	1	0*	38	62*	101
Nonattentive	913	1	2	4	43	50	100
College							
Attentive	417	0	3	3	36*	58*	100
Nonattentive	1005	1	4	2	43	50	100
Total							
Attentive	670	0*	3*	2*	36*	59*	100
Nonattentive	3359	1	5	6	44	44	100

ing that the funds used for science are either a
good investment or not so good an investment. The
item received the lowest overall agreement level,
but provided a very clear differentiation between
the attentive and nonattentive respondents (see
Table 8.4). In all cohorts, the proportion of
"agree" and "strongly agree" responses was significantly higher (at the 0.05 level) for attentives,
and the level of "disagree" responses was significantly higher (at the 0.05 level) for nonattentives.
On the two previous items, the levels of agreement
were significantly different in almost all cases,
but the levels of disagreement tended to be smaller
and not statistically significant. The emergence of
a clearer negative position by the nonattentive respondents makes this item a particularly useful index component.
    The fourth and final item on the factor re-
flects a conclusion that the benefits of science and
technology have outweighed the risks or harms. Ap-

proximately two-thirds of the NPAS respondents indicated a positive view of science and technology on
balance, with about one in five respondents expressing a negative conclusion (see Table 8.5). As with
the previous three items, the proportion of "strongly agree" responses is significantly higher for attentives than nonattentives in all three cohorts,
and there are minimal differences among the cohorts.
In both high school cohorts, the proportion of undecided respondents is particularly high among the
nonattentive respondents, who appear to have given
little thought to the question of the balance of
good and harm.
    When the four items are combined in a Benefits
of Science and Technology Index (BESTI), the difference between attentive and nonattentive respondents
becomes more pronounced. Looking first at the distribution of responses on the BESTI, the proportion

Table 8.4. Distribution of Responses to Item:
"The Federal Government Should Spend More
Money on Science Research."

Group	N	Responses					
		SD	D	U	A	SA	Total
HS-No Coll							
Attentive	56	4%	14%	21%	45%	17%*	101%
Nonattentive	1149	7	21	28	35	10	101
HS-Coll Bnd							
Attentive	185	0*	11*	18	47	24*	100
Nonattentive	910	3	20	21	40	16	100
College							
Attentive	417	3	11*	16*	48*	22*	100
Nonattentive	1003	2	18	24	42	14	100
Total							
Attentive	670	2*	12*	17*	47*	23*	101
Nonattentive	3330	5	19	24	39	13	100

Table 8.5.   Distribution of Responses to Item:
"Overall, Science and Technology Have Caused
More Good Than Harm."

Group	N	SD	D	U	A	SA	Total
HS-No Coll							
Attentive	56	5%	6%*	14%*	55%*	21%	100%
Nonattentive	1161	6	16	25	41	13	101
HS-Coll Bnd							
Attentive	185	5	13	8*	39	35*	100
Nonattentive	913	6	15	15	45	19	100
College							
Attentive	417	7	11	12	44	27*	101
Nonattentive	1005	7	13	15	45	20	100
Total							
Attentive	670	6	11*	11*	43	29*	100
Nonattentive	3359	6	15	19	43	17	100

of high scores is significantly higher (at the 0.05
level) for attentives over nonattentives, and the
proportion of low scores is significantly higher
(at the 0.05 level) for the nonattentives (see Table
8.6). The mean scores on the BESTI are significant
ly higher for the attentives in all three cohorts,
but the differences in the mean scores among the cohorts are not significant. The difference between
the attentives and nonattentives is displayed graph-
ically in Fig. 8.1, which shows the proportion of
attentive and nonattentive respondents scoring three
or more on the BESTI.
     The pattern of responses on both the individual
items in the BESTI and the distribution of scores on
the index, suggests that nearly all of the NPAS respondents hold a generally positive and expectant
view of science and technology. Those respondents
classified as members of the attentive public for
science and technology are significantly more positive toward organized science than those persons

not attentive to science and technology issues, but
there does not appear to be a developmental pattern
in the levels of positive attitude. In view of the
developmental patterns noted earlier in regard to
attentiveness itself, these data would suggest that
the commonalities among attentives outweigh the differences that may be associated with the various cohorts or other groupings.

Attitudes Toward the Risks of Science
and Technology

     The other side of the coin is the perception of
risks involved in the practice of science and technology. In view of the major media attention devoted to nuclear accidents, nuclear waste storage and
disposal, pesticide pollution, food additives, and
chemical toxicity in recent years, it would be reasonable to expect a higher level of awareness of the

Table 8.6. Index of Benefits of Organized Science

Group	N	Index Score					Total	Mean
		0	1	2	3	4		
HS-No Coll								
Attentive	56	0%*	5%*	21%	36%	37%*	99%	0.8
Nonattentive	1161	8	17	26	29	20	100	1.5
HS-Coll Bnd								
Attentive	185	0*	4	17	32	47*	100	0.7
Nonattentive	913	2	11	20	35	31	99	1.1
College								
Attentive	417	2	8*	14	30	47*	101	1.0
Nonattentive	1005	2	12	21	33	33	101	1.1
Total								
Attentive	670	1*	7*	15*	31	46*	100	0.9
Nonattentive	359	4	14	22	32	27	99	1.3

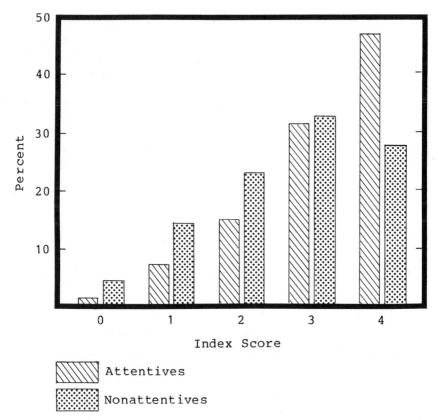

Fig. 8.1.   Distribution of scores on the science
           and technology benefit index.

risks associated with science and technology today
than would have been found a decade ago. The two
principal analysis problems concern the impact of
attentiveness on risk perceptions and the relative
level of risk concerns to benefit perceptions in
both the attentive and the nonattentive publics.
     As in the analysis of benefit perceptions, it
is useful to begin the examination of risk concerns
with a review of the distribution of responses on
each of the four items comprising the Risks of Science and Technology Index (RISTI). After looking at
the individual items, the analysis will turn to the
combined index.
     The strongest loading item on the risk factor
was the statement that science "makes our lives
change too fast." The item is particularly useful
in tapping a general mood since it is almost totally

devoid of any substantive example and relies on the
undifferentiated feeling that things are changing
too fast and that a part of the responsibility for
this situation rests with science. The item was
used in all three of the Science Indicators studies
sponsored by the National Science Board (1973, 1975,
1977) and was very useful in in tapping an adult
disquiet about the impact of science and technology.
Slightly over 60 percent of the attentives and somewhat less than half of the nonattentives disagreed
with the statement, and the difference is significant at the 0.05 level. In this case, there appears
to be a slight developmental pattern, with the high
school attentives (both college-bound and non-
college-bound) being significantly more likely to
disagree with the statement than their nonattentive
classmates, but the difference between attentives
and nonattentives narrows substantially in the college cohort (see Table 8.7). This pattern suggests

Table 8.7. Distribution of Responses on Item:
"One Trouble With Science Is That It Makes
Our Lives Change Too Fast."

				Responses			
Group	N	SD	D	U	A	SA	Total
HS-No Coll							
Attentive	56	12%	48%*	8%*	22%	9%	99%
Nonattentive	1161	5	34	20	32	9	100
HS-Coll Bnd							
Attentive	185	9	55*	5	26	5	100
Nonattentive	913	7	46	12	28	7	100
College							
Attentive	417	12*	49	10	23	6	100
Nonattentive	1005	7	51	11	27	4	100
Total							
Attentive	670	11*	50*	8*	25*	6	100
Nonattentive	3359	7	43	15	29	7	101

that the college experience may have the effect of
enhancing the awareness of risks for the attentives,
bringing them closer to the somewhat more doubtful
nonattentives.
       The second strongest loading item on the risk
factor was a statement that the growth of science
"means that a few people could control our lives."
Like the previous statement, it is a somewhat vague
and general statement designed to tap an underlying
mood about science and technology rather than a specific policy-related point. Unlike the first item,
this statement does not show a clear difference between attentives and nonattentives in the levels of
either agreement or disagreement (see Table 8.8).
The data do suggest a slight developmental pattern
among the nonattentives, with the proportion of "uncertain" responses dropping noticeably between high
school and college, and disagreement growing with
increased educational expectations or experience, so
that the attentive and nonattentive respondents in
the college cohort are almost indistinguishable.
       The third strongest loading item on the risk
factor was the statement that science tends to
break down "people's ideas of right and wrong." The
item is only slightly more specific than the first
two, referencing moral questions in a nonreligious
framework. This item has also been used in the National Science Board studies and was useful in those
analyses. In the NPAS, the item provided a clear
differentiation between the college-bound and non-
college-bound students, but little beyond that division (see Table 8.9). Among the non-college-bound
high school cohort, both attentives and nonattentives were slightly more likely to agree with the
statement than disagree with it, and in almost equal
proportions. In contrast, a clear majority of college-bound high school students disagreed with the
statement and a plurality of nonattentives in that
cohort disagreed with it. Among the college cohort,
a full 75 percent of attentives and 64 percent of
nonattentives disagreed with the statement. Overall,
it would appear that concern about science as a
source of moral erosion is strongest among the non-
college-bound cohort who have had the least exposure
to science instruction and who will have low levels
of exposure to science during their lifetimes, although they may be exposed to a good deal of technology, as LaPorte (1978) points out.
       The weakest of the four items that loaded on
the risk factor was a statement asserting that "we

Table 8.8. Distribution of Responses on Item:
"The Growth of Science Means That a Few
People Could Control Our Lives."

		Responses					
Group	N	SD	D	U	A	SA	Total
HS-No Coll							
Attentive	56	17%	34%	16%*	33%	1%*	101%
Nonattentive	1161	12	32	26	25	6	101
HS-Coll Bnd							
Attentive	185	12	48*	13*	23	4	100
Nonattentive	913	15	37	21	21	5	99
College							
Attentive	417	13	51*	12*	18	6	100
Nonattentive	1005	13	45	17	22	4	101
Total							
Attentive	670	13	49*	12*	21	5	100
Nonattentive	3359	13	38	21	23	5	100

depend too much on science and not enough on faith,"
which served to focus directly on the possibility of
a science-religion conflict. While scientists and
theologians have largely made their peace in modern
industrial societies, the National Science Board
studies and other studies have found lingering concern on this perceived conflict. The general distribution of responses on the item followed the pattern of the preceding item on right and wrong, but
produced a clearer differentiation among the three
cohorts (see Table 8.10). Among the non-college-
bound high school cohort, attentives were slightly
more likely to disagree with the statement than
argee with it, but nonattentives were more likely to
agree with it than reject it. In the college-
bound high school cohort, a majority of both attentives and nonattentives rejected the statement, but
the attentives were significantly (at the 0.05
level) more opposed to the statement than were the
nonattentives. Among the college cohort, a full two-

thirds of both the attentives and the nonattentives
rejected the statement. This pattern suggests, generalizing from cross-sectional data, that the nonattentive segment of the NPAS population experienced a
substantial attitude change during the high school
and college years, resulting in an attitude profile
not significantly different from that of the attentives in the college cohort.
    To assess the combined impact of the four items,
a Risks of Science and Technology Index was calculated, giving a positive score of one each time a
respondent agreed or agreed strongly with an item.
The RISTI ranges from zero to four and the distribution of the NPAS respondents on the index is displayed in Table 8.11. Following the pattern found in
the individual items, the RISTI shows no significant
differences between the attentives and the nonattentives within any of the three cohort groupings.
Overall, the data indicate a declining concern about

Table 8.9. Distribution of Responses on Item:
"One of the Bad Effects of Science is That it
Breaks Down People's Ideas of Right and Wrong."

Group	N	SD	D	U	A	SA	Total
HS-No Coll							
Attentive	56	8%	22%	28%	37%	5%	100%
Nonattentive	1161	5	26	28	35	6	100
HS-Coll Bnd							
Attentive	185	14	40	18*	21	7	100
Nonattentive	913	9	34	26	27	4	100
College							
Attentive	417	27*	48	13*	12	1	101
Nonattentive	1005	20	44	23	11	2	100
Total							
Attentive	670	21*	43*	16*	17*	3	100
Nonattentive	3359	11	34	26	25	4	100

risk over the three cohorts, with the highest level
of risk concern expressed by the non-college-bound
cohort and the lowest level being registered by the
college cohort. In contrast to the BESTI, the differences in the mean scores of the RISTI are not
significant between attentives and nonattentives,
but are significantly different among the three cohorts. The absence of a meaningful difference is
demonstrated in the graphical comparison of the attentives and nonattentives on RISTI in Fig. 8.2.

The Balance of Risks and Benefits

In the preceding two sections, the perceptions
of the NPAS respondents toward the benefits and
risks of science and technology have been examined

Table 8.10. Distribution of Responses on Item:
"One of Our Big Troubles is That We Depend Too
Much on Science and Not Enough on Faith."

		Responses					
Group	N	SD	D	U	A	SA	Total
HS-No Coll							
Attentive	56	21%*	27%	16%	27%	9%	100%
Nonattentive	1161	8	29	18	32	12	99
HS-Coll Bnd							
Attentive	185	21*	47*	13	14*	5*	100
Nonattentive	913	13	38	15	23	12	101
College							
Attentive	417	23*	45	11*	15	5	99
Nonattentive	1005	18	49	15	14	4	100
Total							
Attentive	670	22*	44*	12*	16*	6*	100
Nonattentive	3359	12	38	16	24	10	100

separately. Since concrete policy views most often
reflect some balancing of risk and benefit, it is
important to inquire into the balance between perceptions of risk and perceptions of benefits and the
impact, if any, of attentiveness on the nature of
the balance.

On the conceptual level, it is possible to
think about the combinations of risk and benefit
perceptions and to identify four types of respondents (see Fig. 8.3). If a respondent viewed science and technology as possessing high benefits and
low risk, the person might be appropriately labeled
an Advocate. In contrast, if a respondent expressed
a high level of concern about risks and a low level
of expectation about benefits, it would be appropriate to label that type of respondent a Doubter. If
a respondent expressed high levels of both benefit
expectation and risk concern, it would be appropriate to label the person a Balancer. And, if a person expressed low levels of both benefit and risk

Table 8.11. Index of Risks of Organized Science

| Group | N | Index Score | | | | | Total | Mean |
		0	1	2	3	4		
HS-No Coll								
Attentive	56	27%	35%	17%	9%*	12%	100%	1.7
Nonattentive	1161	25	26	24	18	8	101	1.6
HS-Coll Bnd								
Attentive	185	40	28	20	9	2	99	1.2
Nonattentive	913	34	27	22	12	4	99	1.4
College								
Attentive	417	47	31	15	6	2	101	1.0
Nonattentive	1005	46	29	16	7	2	100	1.0
Total								
Attentive	670	43*	30	18*	7*	3*	101	1.1
Nonattentive	359	34	27	21	13	5	100	1.4

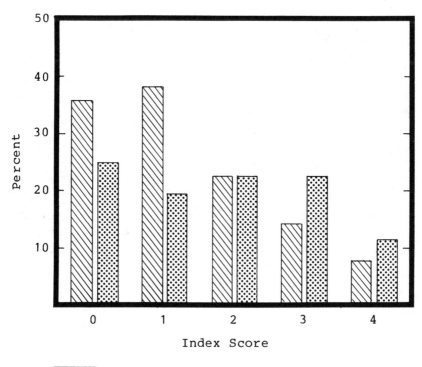

Fig. 8.2.   Distribution of scores on the science
and technology risk index.

attitudes, it is likely that the respondent has little interest in science and technology and might be
most appropriately termed a Neutral.

Using these four types, the NPAS data indicate
that attentives are most likely to be Advocates and
that this tendency is strongest among the college
cohort (see Table 8.12). A similar developmental
pattern is implied for the nonattentives, but the
proportion of nonattentives who are Advocates is
significantly less than that of attentives in all
three cohorts. As noted in the preceding discussions of the separate indices of risk and benefits,
the 1978 NPAS data suggest that the effect of college is to enhance the perception of benefits and
reduce concern about risks, thus reducing the proportion of Balancers and increasing the proportion

of Advocates. It is also interesting to note that
the nonattentive non-college-bound high school cohort is almost equally split among the four types.
Since this group will constitute approximately half
of the adult electorate in the future, the presence
of large segments of Neutrals and Doubters in this
group should be noted.

   Assuming that the BESTI and the RISTI measure
two facets of a more general attitude toward science
and technology, another approach to the development
of a composite measure is a combination of the two
sets of items, giving a positive score to agreement
with the benefit items and a positive score for disagreement with the risk items. This procedure would
produce an Index of Support for Science and Technology, with a range from zero to eight. This proce-

Risk

	Low	High
Benefit Low	Neutrals	Doubters
Benefit High	Advocates	Balancers

Fig. 8.3.   Risk-benefit typology.

Table 8.12.   Structure of Risk and
Benefits Attitudes

Group	N	Low R Low B	Low R High B	High R Low B	High R High B	Total
HS-No Coll						
Attentive	56	17%*	62%	9%	12%	100%
Nonattentive	1161	38	37	13	12	100
HS-Coll Bnd						
Attentive	185	18*	70*	2*	9	99
Nonattentive	913	27	57	7	10	101
College						
Attentive	417	21*	71*	3	5	100
Nonattentive	1005	30	62	4	4	100
Total						
Attentive	670	20*	70*	3*	7*	100
Nonattentive	3359	33	50	9	9	101

dure would place the Advocates near the top of the
index and place the Doubters near the bottom of the
index. The Balancers and the Neutral would both be
clustered in the middle of the index.
     For analysis purposes, the procedure outlined
above was followed and a General Attitude toward
Science and Technology Index (GASTI) was formed. The
scores were grouped into high, medium, and low categories; the distribution of the scores on the GASTI
indicates that attentives in all cohorts are more
likely to score high on GASTI and less likely to
score low on GASTI than their nonattentive counterparts (see Table 8.13). The data suggest a slight
developmental pattern, with the proportion of both
attentives and non-attentives scoring high on the
GASTI increasing with each cohort, from the non-
college-bound high school group to the college student cohort.
     In summary, looking at the general attitudes of
the NPAS repondents toward science and technology in
American society, the data indicate that these young

Table 8.13. Distribution of Scores on the
General Attitude Toward Science and
Technology Index (GASTI)

| Group | N | Index Score | | | Total |
		Low 0-2	Medium 3-5	High 6-8	
HS-No Coll					
Attentive	56	9%*	51%	40%*	100%
Nonattentive	1161	26	51	23	100
HS-Coll Bnd					
Attentive	185	2*	42	56*	100
Nonattentive	913	13	49	39	101
College					
Attentive	417	7	28*	65*	100
Nonattentive	1005	10	39	51	100
Total					
Attentive	670	6*	34*	60*	100
Nonattentive	3359	17	47	36	100

Americans display a generally positive outlook, holding high expectations for science and technology.
At the same time, a significant portion of the respondents evidenced a moderate to high level of concern about the potential risks of science and technology. In almost every grouping of items and cohorts, those respondents classified as attentive to
science and technology issues were more positive toward science and technology and more willing to take
certain risks in the pursuit of the possible benefits of science and technology. Recalling the mon-
itoring function of the attentives in the political
system, the 1978 NPAS data would suggest that the
new entrants into the attentive public for science
and technology will hold the scientific community to
high expectations for achievements and will be willing to take some risks in the pursuit of those objectives.

SPECIFIC SCIENCE AND TECHNOLOGY POLICY VIEWS

In addition to the more general positive and
negative attitudes toward science and technology examined in the preceding section, individuals also
have viewpoints on specific policy issues that may
be related to science and technology. As was
observed in chapter one, the number of science and
technology related issues on the national political
agenda has increased markedly in recent years and
promises to grow in future years. It is important
to understand the present policy views of young
Americans on a representative set of science and
technology issues and to explore the impact of attentiveness to organized science on specific issues.
For this purpose, this section will focus first on
a set of specific energy policy issues, then on the
international competition aspects of science and
technology, and finally on regulatory issues and the
exploration of outer space. The section will conclude with an analysis of the federal spending priorities of the respondents, and will explore the
role of attentiveness in spending priorities.
     At the present time and for the foreseeable fu-
ture, the issues surrounding the energy problem will
be high on the national agenda, and several of those
issues involve scientific and technological issues.
Perhaps the most controversial of those issues is
the question of the safety and utility of nuclear
power for the generation of electricity. The 1978
NPAS questionnaire included several items about nuclear power and other sources of energy and those
data provide an excellent starting point for an
analysis of young adult attitudes and the impact of
attentiveness on these specific policy views. It
should be noted that the NPAS was conducted almost
a year before the Three Mile Island accident.
     The NPAS respondents were presented with the
statement that "there are still major unanswered
questions about the safety of nuclear power plants,"
and asked to express strong agreement, agreement,
disagreement, strong disagreement, or uncertainty.
The data indicate that approximately 80 percent of
the NPAS respondents have some reservations about
the safety of nuclear power, and there were no significant differences between attentives and nonattentives in any of the three cohorts (see Table

Table 8.14.   Attitudes Toward Selected
Energy Policy Issues

Group	N	Percent Agreeing With Item...			
		A	B	C	D
HS-No Coll					
Attentive	56	81%	37%	66%	61%*
Nonattentive	1161	82	28	59	47
HS-Coll Bnd					
Attentive	185	88	30	65	64*
Nonattentive	913	86	25	62	55
College					
Attentive	417	80	28	56	68*
Nonattentive	1005	84	21	56	56
Total					
Attentive	670	82	30*	60	66*
Nonattentive	3330	83	25	59	53

Item A: There are still major unanswered questions
about the safety of nuclear power plants.
Item B: The risk involved in generating nuclear
power is relatively minor and should not
block the construction of new nuclear power
plants.
Item C: Solar energy is the best single long-term
solution to our energy problem.
Item D: We can depend on science and technology for
a long-term solution to the energy problem.

8.14). To provide a check on the directionality of
the question, a second item was included in the survey, asking respondents to indicate their agreement
or disagreement with the statement, "The risk involved in generating nuclear power is relatively
minor and should not block the construction of new
nuclear power plants." The data show that only
about 30 percent of the NPAS respondents would agree
with the statement, but that the attentives in all
three cohorts were significantly (at the 0.05 level)
more likely to agree with the statement than were
the nonattentives (see Table 8.14). Overall, the
data suggest a weak developmental pattern, with the
non-college-bound high school cohort being the most
likely to agree with the statement and the college
cohort being the least likely to agree with the
statement. This pattern indicates the level of
consistency grows with increased formal educational exposure.

Many of the strongest critics of nuclear power
have suggested that solar power should be utilized
as the long-term solution for the energy problem,
and the 1978 NPAS included the statement that "solar
energy is the best single long-term solution to our
energy problem." An analysis of the responses to
the item indicates that a majority of young Americans agree with the statement, but that attentives
in the college-bound high school and the college cohorts are no more likely to endorse the statement
than their nonattentive colleagues (see Table 8.14).
The college cohort is significantly less likely to
agree with the statement than either of the high
school cohorts. Viewed in the context of the previous items, these data would suggest that the students in the college cohort have serious doubts
about the use of nuclear power and that a majority
are willing to look to solar power as the best alternative.

While a number of the respondents appear to be
doubtful of both nuclear and solar energy sources,
there is a high level of agreement with the proposition that science and technology can be depended
upon to find a long-term solution to the current
energy difficulties. In all three cohorts, attentives were significantly (at the 0.05 level) more
likely to express this view than were the nonattentive respondents, with approximately two-thirds of
the attentives endorsing the proposition (see Table
8.14). The nonattentive members of the non-college-
bound high school cohort were the least likely to

express agreement with the statement, with only 47
percent agreeing or strongly disagreeing.

In summary, the 1978 NPAS data indicate that
there are some specific policy differences in the
energy area between attentives and nonattentives,
and that there are also substantial areas of agreement between the two groups, especially in the college cohort. The two areas of disagreement are
important in both real and symbolic terms. The major differences between the attentives and the nonattentives appear to revolve around the willingness
to build more nuclear power plants and the expectation that science and technology will provide a
long-term solution for the energy problem. The attentive group's willingness to support more nuclear
power facilities illustrates the greater willingness of the attentives to take a calculated risk,
and the higher level of expectations for a long-
term solution from science and technology reflects
the more general expectation of benefits discussed
in the previous section. The willingness to take
some risk and an underlying high level of faith and
expectation in science and technolology are appropriate symbols of the general outlook of the attentive public for organized science.

The 1978 NPAS instrument also explored the policy thinking of the respondents in a number of areas
other than energy. Two items focused on the international competition aspect of science and technology. The first statement asserted that it was
important for the United States to "continue to develop new weapons at least as fast as the Russians,"
tapping both the international competition aspect
and the Cold War aspects of the respondent's thinking. In all cohorts, the attentives were more
likely to agree with the statement than their nonattentive counterparts, and the data show a cohort effect with the level of agreement being significantly
(at the 0.05 level) lower among the college cohort
(see Table 8.15). These data are particularly interesting in that they suggest the absence of over-
arching belief systems. In regard to nuclear power,
the NPAS respondents tended to take the more liberal
position and by fairly substantial majorities, but
in the weapons area, a majority of all cohorts took
what would generally be viewed as the more conservative or hawkish policy position.

Table 8.15. Attitudes on Selected Science
and Technology Policy Issues

Group	N	Percent Agreeing with Item...			
		A	B	C	D
HS-No Coll					
Attentive	56	74%*	18%	36%	3%*
Nonattentive	1161	62	22	44	16
HS-Coll Bnd					
Attentive	185	73	22	28*	4
Nonattentive	913	66	20	42	8
College					
Attentive	417	52	10	19*	4
Nonattentive	1005	47	13	30	4
Total					
Attentive	670	60	14*	11*	4*
Nonattentive	3330	59	19	39	10

Item A: It is important that the U.S. continue
to develop new weapons at least as fast as
the Russians.

Item B: The U.S. has lost its lead in science
research to the Soviet Union and other
nations.

Item C: When all is said and done, the space program still hasn't produced much of value
for this country.

Item D: A privately-owned chemical company should be
allowed to manufacture and sell anything it
wants without governmental interference.

A second statement concerning international
competition declared that the United States "has
lost its lead in science research to the Soviet
Union and other nations." This statement is a good
reflection of a major argument for increased science
and technology activities that was used extensively
during the years immediately after Sputnik. The
1978 NPAS data indicate that it is not a popular
position among today's young Americans, with fewer
than one in five of the respondents agreeing with
the statement. The proportions of attentives and
nonattentives agreeing with the statement are not
significantly different, but the proportion of
agreement in the college cohort is significantly
lower (at the 0.05 level) than either of the high
school cohorts (see Table 8.15). From these data,
it would appear that the motivation to support more
science research does not come from a sense of international inferiority in science, although the
support for continued weapons development may be
spurred in part by competitive attitudes.

A third statement concerned the space program,
and asserted that "the space program still hasn't
produced much of value for this country." The
statement was worded in the negative to provide some
balance in the directionality in the items in the
instrument, and not as a reflection of any particular policy position. Since the launching of Sputnik by the Soviet Union in 1957 marked the beginning
of a national drive in science and technology research and in science and engineering education, it
is appropriate to look at the policy views of the
newest generation of Americans about the value of
the space program. In this regard, it is useful to
recall that the vast majority of the 1978 NPAS respondents were born after 1957 and have experienced
space exploration throughout their lives. Unlike
older adults who have clear memories of the shock of
Sputnik and the mobilization of the education establishment to compete in a space race, the young
adults in the NPAS may be expected to see the activity as normal.

The distribution of responses on the space
program item indicates that attentives are significantly less likely to be critical of the space
program, especially among the college-bound high
school cohort and the college cohort (see Table
8.15). The level of program criticism among the
college cohort is significantly (at the 0.05 level)

lower than either of the two high school cohorts.
This result suggests a generally high level of
support.

A final specific policy item focused on the
regulation of science and technology activities; a
statement was used that declared that "a privately
owned chemical company should be allowed to manufacture and sell anything they want without governmental interference." This is a somewhat doctri-
naire viewpoint, but it was hoped that it would tap
the respondents' views about the appropriateness of
regulations, particularly as they relate to the application of science in specific technologies. The
data indicate a strong support for the regulation
of chemical companies in all cohorts and no distinguishable pattern of differences between attentives
and nonattentives (see Table 8.15). Given the extreme position of the statement, the results should
not be interpreted as an endorsement of all regulations 'present and future' but rather as an endorsement of the right of a public body to impose regulations on private firms involved in scientific and
technological activities.

Another approach to understanding the specific
policy views of the NPAS respondents utilizes the
vehicle of asking for a selection of spending priorities in science and technology. Using the idea
of an allocation of federal expenditures for various
scientific and technological   activities, a question on the 1978 NPAS asked the respondents to select from a list of 13 activities the three types of
research for which they would most like to see their
federal tax dollars used and the three types of research for which they would least like to see their
federal tax dollars used. This approach is not in-
tended to be a serious judgment about alternative
federal budgetary choices, but rather a barometer
of the importance attached to the various areas of
scientific and technological work now supported by
federal funding.

The data from the NPAS indicate that members of
the attentive public for science and technology differ significantly from their nonattentive counterparts on their assessment of the relative importance
of a number of types of scientific and technological
research (see Table 8.16). Members of the attentive
public are significantly (at the 0.05 level) more
likely than their nonattentive colleagues to support energy research, weapons development, and space
exploration, and they are significantly less likely

Table 8.16. Federal Spending Priorities for
Science and Technology

		Percent Giving Favorable Response on...						
Group	N	A	B	C	D	E	F	G
HS–No Coll								
Attentive	45	47%	44%*	40%	23%	23%	35%	15%
Nonattentive	982	33	59	34	29	27	35	18
HS–Coll Bnd								
Attentive	170	61*	40*	34	41	23	16*	13
Nonattentive	814	45	54	33	41	27	22	14
College								
Attentive	395	70*	32	49	35*	35	9	
Nonattentive	964	64	30	48	47	38	11	16
Total								
Attentive	610	66*	35*	44*	36	31	13*	13
Nonattentive	980	46	48	38	39	31	23	16

Item A: Developing new energy sources.
Item B: Reducing crime.
Item C: Reducing and controlling pollution.
Item D: Improving education.
Item E: Improving health care.
Item F: Improving the safety of automobiles.
Item G: Finding better birth control methods.

Table 8.16 (continued).   Federal Spending Priorities
            for Science and Technology

Group	N	Percent Giving Favorable Response on...					
		H	I	J	K	L	M
HS-No Coll							
Attentive	45	30%*	8%*	21%*	9%	5%	3%
Nonattentive	982	15	21	8	8	9	4
HS-Coll Bnd							
Attentive	170	25	7*	20*	10	5	4
Nonattentive	814	20	15	12	7	5	4
College							
Attentive	395	13*	4	18*	9	12	2
Nonattentive	964	8	6	12	10	10	2
Total							
Attentive	610	18*	5*	19*	9	10	2
Nonattentive	2980	14	14	10	9	8	3

Item H: Developing or improving weapons for
         national defense.
Item I: Finding new methods for preventing and
         treating drug addiction.
Item J: Space exploration.
Item K: Discovering new basic knowledge about man
         and nature.
Item L: Developing faster and safer public trans-
         portation.
Item M: Weather control and prediction.

to support research to reduce crime, improve auto-
motive safety, or cure drug addiction. The pattern
suggests that nonattentives are more likely to favor
expenditures that relate directly to individuals,
while the attentives tend to place higher priorities
on broader and less personal research objectives.
This is a substantial difference in the policy perspectives of the two groups, and would lend support
to LaPorte's contention that nonattentives tend to
experience only technology and to think about science in technological terms. The high priority
placed on auto safety, drug addiction cures, and
crime reduction or prevention illustrates the narrower and more personal policy perspective of the
nonattentive, or general, public.

SUMMARY

    Looking at both general and specific policy
views, it appears that the attentive public for science and technology holds higher expectations for
organized science than the nonattentives, but are no
less aware of potential risks. Attentives are most
likely to be Advocates of organized science, using
the terminology of the typology proposed above, or
Balancers of the risks and benefits.
    These general attitudes toward science and
technology carry forward in large part to more specific policy attitudes. For example, the attentives are significantly more likely to favor the
construction of additional nuclear power plants or
to expect a long-term energy solution from science
and technology than the nonattentives. The attentives are more likely to support space exploration
and weapons development than nonattentives and
less likely to endorse programs primarily of personal or individual benefit.
    The original inquiry in this chapter can now be
addressed: it appears that being a member of the
attentive public for organized science does make a
difference in both general and specific policy perspectives. Since the general literature on political participation would suggest that the views of
the attentives are more likely to be transmitted
into the policy formation process, it is and will
continue to be important to understand the structure, the composition, and the views of the attentive public for organized science.

IV

The Sources of Attentiveness

9 The Family

Families influence the development of attentiveness
to organized science in at least three ways. First,
the family provides children with a social status
and an identity within an ongoing community. Birth
into a particular family determines, to some extent,
the quality of the schools one attends, the religious institutions one is exposed to, and the rela-
tives and "friends of the family" one interacts with.
In addition, the educational and occupational background of each member influences the educational and
occupational environment children experience as they
grow up. Homes that contain a large number of books
or an abundance of tools and gadgets foster different attitudes and abilities than homes that lack
these resources for learning.
    A second way the family influences the develop-
ment of attentiveness is through example. When
people have a particular interest, they tend to involve themselves, and often their children, in activities related to that interest. Parents who are
interested in science and technology, for instance,
may subscribe to magazines and newspapers that cover
scientific issues or they may watch science documen-
taries on television, take nature walks, visit science museums and zoos, or attend lectures on scientific topics. In being attentive to science issues
themselves, parents set an example of attentiveness
for their children. If they also make an effort to
involve their children in these activities and encourage their children to discuss what they learn,
parents can foster the habit of exploration in general and attentiveness to science in particular.

A third way the family influences the development of attentiveness is through directly teaching
interests and skills. This teaching, however, need
not represent a direct transmission of the parents'
own background and interest. Some parents recognize
that the world in which they grew up is not the
world in which their children will live. In a rapidly changing scientific and technological environment these parents might take a special interest in
their children becoming aware of basic scientific
concepts and interested in major technological break-
throughs (cf. Maccoby 1968; Inkeles 1955). To the
extent that members of a family make an effort to
teach children the rudiments of scientific understanding, families may contribute to the development
of attentiveness as the children grow older.

SOCIOECONOMIC STATUS

Two components of social status that have consistently been shown to correlate with attentiveness
to public issues are educational achievement and
occupation (see e.g., Berelson et al. 1954; Campbell et al. 1960). People with higher levels of
education are, in general, more likely to maintain
interest in abstract public issues, to keep informed
about current events relevant to these interests,
and to participate in the political process to affect
changes in public policy related to these issues
(Milbrath 1965). Likewise, people who share the
same occupation frequently share similar political
interests and views (see review by Sigel & Hoskin
1977). It is not surprising, then, to find that
these same factors have been shown to affect the
child-rearing practices of parents (Kohn 1969;
Bernstein 1971) and the occupational and educational
aspirations parents hold for their children (Boocock
1972; Sewell et al. 1976; Duncan et al. 1972; Gordon
1971). Both directly through the attitudes parents
stress and indirectly through the educational and
occupational environment they provide, parents of
different socioeconomic statuses tend to raise children with different political views and different
propensities to participate in the political process
(Marvick & Nixon 1961; Flacks 1967; Chaffee et al.
1977).
To determine the relationship between different
educational and occupational environments and the

development of attentiveness to science, each student was asked to report the level of education completed by his or her parents and their occupations.
The highest level of education reported for either
parent was used as a measure of the educational environment of the student's family.(1) The highest
prestige of either parent's occupation (as rated on
the Hodge, Siegel, Rossi [1964] occupational prestige
scale) was used as a measure of the occupational
component of each family's social status.(2)
     Four levels of family education were distin-
quished. Twelve percent of the students reported
that neither parent in their family completed high
school, while 51 percent indicated that at least
one parent completed high school, junior college,
or a vocational degree. Another 19 percent of the
students reported that at least one parent had completed a bachelor's level degree, while the remaining 18 percent reported that at least one parent
had completed a graduate or professional degree.
These four levels of family education will be used
to examine whether the level of education completed by parents has a discernible effect on the development of attentiveness to organized science
among children in the family.
     Three levels of occupational prestige were dis-
tinquished. Forty-six percent of the students in
the NPAS sample reported that one or both of their
parents are employed in semiskilled occupations or
the skilled trades. These blue-collar occupations
are consistently rated as relatively low on occupational prestige by the general public (score 40 or
less on the Hodge, Siegel, Rossi scale). Forty-two
percent of the students reported that at least one
parent is employed in one of the technical occupations, in business sales, in farming, or in one of
the lower prestige professions (e.g., nursing, accounting, engineering, elementary or secondary
teaching). These mostly white-collar occupations
are consistently rated in the middle range of occupational prestige (43 to 69 on the Hodge, Siegel,
Rossi scale). The remaining 12 percent reported at
least one parent is a business owner or executive or
practices one of the higher status professions (e.g.,
law, medicine, college teaching). These business
and professional occupations are consistently rated
high on occupational prestige (70 to 82 on the Hodge
Siegel, Rossi scale).
     The cross-classification of parents' education
and parents' occupational prestige defines 12 dis-

tinct categories of family socioeconomic status (see
Table 9.1). The general pattern revealed by the
data is that the higher the social status of a student's family, the greater the likelihood that the
student is attentive to science. The differences,
however, appear to be due exclusively to the educational component of family status. High school and
college students whose parents have completed four
or more years of college are 10 to 15 percent more
likely to be attentive to science than are students
whose parents have not completed college. Among families with comparable levels of education, the prestige of the parents' occupations appears to make no
systematic difference in the likelihood that children in the home will become attentive to organized
science during their high school or college years.
    The absence of any effect of parents' occupa-
tional prestige on their children's attentiveness to
science is interesting in the light of considerable
research that has shown that adults in different occupations differ considerably in their attitudes and
attentiveness to public issues (see e.g., Kornhauser
1965; Toren 1972; Ewing 1964; Glaser 1960 1964; Ladd
and Lipset 1972; Ziegler 1966) and that parents who
work in different kinds of occupational settings develop different political values that they seek to
pass on to their children (Miller & Swanson 1958;
Kohn 1969; Bernstein 1971). The present research
suggests, however, that the impact of parents' own
attentiveness to science on the development of attentiveness in their children varies primarily as a
function of the educational component of social
status and not as a function of systematic differences in occupational prestige.

EDUCATIONAL ASPIRATIONS

The finding that attentiveness to organized science
correlates with parents' level of education suggests
that the educational environment of the family is an
important factor contributing to the development
of attentiveness. How does this influence work?
One way may be through provision of opportunities
and encouragement for children to become interested
in science at an early age and to develop aspirations
to learn more about science as they grow older. For
high school students, these aspirations might be reflected in a desire to attend college, while for col-

Table 9.1. Percent Attentive by Parents'
Occupational Prestige and Parents' Education

Group	Parents' Education	Parents' Prestige	Attentive to Science	
			N	%
High School	Less than High School	Low	212	3
		Medium	37	3
		High	13	(1)
	High School/ Vocational	Low	582	9
		Medium	278	9
		High	76	9
	Bachelor's	Low	82	18
		Medium	174	24
		High	36	23
	Graduate/ Profes- sional	Low	61	22
		Medium	178	14
		High	34	24
College	Less than High School	Low	75	31
		Medium	5	(1)
		High	6	(4)
	High School/ Vocational	Low	298	25
		Medium	223	27
		High	102	26
	Bachelor's	Low	46	24
		Medium	197	31
		High	67	40
	Graduate/ Profes- sional	Low	21	38
		Medium	214	36
		High	52	41
Total			3068	20

Notes: The number attentive, not the percent,
is given for cells with fewer than 20
respondents.

Six-hundred-and-sixty-nine respondents'
parents could not be estimated on both
education and occupation.

lege students, these aspirations might be reflected
in the desire to do graduate or professional work.
The NPAS data are consistent with this pattern.
Forty-six percent of the college students in the
sample come from families with higher status.
Clearly, children raised in higher status families
are more likely to attend college, despite the extensive financial resources that have been made
available in recent years to students whose families
cannot afford to send them to college.(4)

Since family social status is related to both
college aspirations and actual college attendance,
it is reasonable to ask whether the combination of
family social status and educational aspirations account for the association between family status and
attentiveness to science. The data indicate that
educational aspirations are associated with family
status and that both family status and educational
aspirations are associated with the development of
attentiveness (see Table 9.2). The average difference in proportion attentive is only 6 percent between students from higher and lower status families
who are in the same educational cohort and share the
same educational aspirations, compared to a 14 percent difference between students from higher and lower status families as a whole.

The logit model that fits these data indicates
that educational aspirations and college attendance
account for some, but not all, of the relationship
between family social status and attentiveness. Controlling for both educational cohort and educational
aspirations, there is still a significant relationship between family status and attentiveness to organized science (see Table 9.3). The model indicates
that family social status has both a direct effect
on the development of attentiveness to organized
science and an indirect effect through educational
aspirations. The effects of family status, educational aspirations, and college attendance on attentiveness to science are additive. The cumulative
effect is that 39 percent of the college students
from higher status families who aspire to graduate
or professional school are attentive to science and
technology compared to 4 percent among high school
students from lower status families who do not expect to complete a bachelor's degree (see Fig. 9.1).

In the analyses that follow, educational aspirations and school cohort will be combined to form a
single "school cohort" variable describing types of
students with differing educational aspirations. The

Table 9.2. Percent Attentive by Students'
Educational Aspirations and Family
Social Status

Group	Social Status	Educational Aspiration	Attentive to Science	
			N	%
High School	Low	HS/Voc/AA	1057	4
		Bachelor's	273	13
		Grad/Prof	426	14
	High	HS/Voc/AA	160	8
		Bachelor's	143	17
		Grad/Prof	256	26
College	Low	HS/Voc/AA	9	(0)
		Bachelor's	252	22
		Grad/Prof	459	30
	High	HS/Voc/AA	8	(2)
		Bachelor's	218	25
		Grad/Prof	406	39
Total			3667	20

Notes: Seventeen college students indicated
they expected to complete a vocational, technical, or junior college
degree.

Seventy college students did not
report their educational aspirations.
Two-hundred ninety-two high school
students who similarly did not report
their educational aspirations were not
included in any of the analyses in
earlier chapters since they could not
be classified as included in one of
the two high school cohorts. These
students have also been removed from
the present analysis.

Table 9.3.   Model for Data in Table 9.2

Models	LRX^2	df	CMPD
H1 ESC,A	366.01	11	--
H2 ESC,AC,AS,AE	7.80	7	.98
H3 H2 minus AC	61.25	8	--
Difference due to AC	53.46	1	.15
H4 H2 minus AS	32.30	8	--
Difference due to AS	24.51	1	.07
H5 H2 minus AE	105.61	9	--
Difference due to AE	97.81	2	.27

A = attentiveness to science
E = educational expectations
S = socioeconomic status
C = educational cohort

variable is trichotomous -- high school-non-college-bound (HSNC), high school-college-bound (HSCB), and college (COLLEGE). This three-category typology will allow an analysis of attentiveness to organized science in three relatively homogeneous subpopulations first described in chapter five.

OCCUPATIONAL ASPIRATIONS

In a manner parallel to the transmission of educational aspirations, families also transmit a set of values about various occupational aspirations for young persons in the household. Traditionally, children, especially males, were expected to follow routinely the occupation of their fathers. One of the important aspects of urbanization and industrialization is the emergence of wider discretionary room for young people to make career choices. The career education movement of the last decade is both a result and a reflection of this change in generational transmission of occupational values and aspirations.

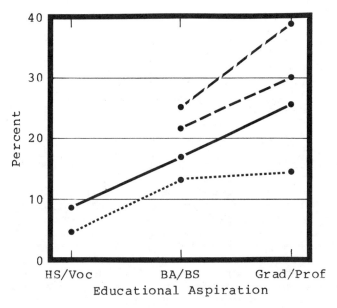

Note: High school/vocational given as an aspiration
        by too few college students for valuable
        estimation of percentage attentive.

●– – – –● Coll High status   ●————● HS High status
●·– – –·● Coll Low status   ●············● HS Low status

Fig. 9.1.   Percentage attentive to organized science
by family social status, educational cohort, and
educational aspirations.

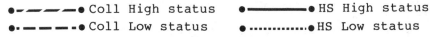

        Since higher levels of educational aspirations
are associated with attentiveness to organized science, it is reasonable to expect that more presti-
gious aspirations will also be associated with attentiveness. Are there separate influences of each of
these distinguishable aspirations on the development
of attentiveness, or does one kind of aspiration
subsume the influence of the other?
        To examine this question, respondents were ask-
ed to report their anticipated occupation and how
sure they were of this choice: very, fairly, or not
at all sure. The anticipated occupations were scored
on the Hodge, Siegel, Rossi occupational prestige
scale and were categorized by a procedure similar to
that used in regard to parental occupations. Students who aspired to occupations with prestige

scores lower than 69 were classified as having medium to low occupational aspirations, while students
who aspired to occupations with prestige scores of
70 to 82 were classified as having higher aspirations. Eighty percent of the students reported occupations they "expected" to work in.
     To build an index of occupational aspirations,
the sample was divided into three groups. Students
who reported occupational plans and who indicated
they were "fairly sure" or "very sure" of their
choice were classified on the basis of the prestige
of their chosen occupation. Students who indicated
they were "not at all sure" of their occupational
choice or who indicated they were "fairly sure" but
did not indicate any particular occupation were classified as being undecided about their occupations.
     The NPAS data indicate that the association of
occupational aspirations with attentiveness to science is different among high school and college students (see Table 9.4). Fig. 9.2 illustrates this
interaction summed across levels of family status.
(See Table 9.5 for terms necessary to fit these
data). In high school, students who have decided on
an occupation are about 5 percent more likely to be
attentive to science than are students who are "undecided" about their occupational choice. The prestige of the occupation the high school students have
chosen for themselves does not appear to make any
difference in whether or not they are attentive to
science. In college, however, the prestige of the
students' chosen occupations does appear to make a
difference. College students who have chosen a
medium or low prestige occupation are 7 percent less
likely to be attentive to science than are students
who are undecided about an occupation and 11 percent
less likely to be attentive than those who have chosen a higher status (business or professional) occupation as their goal.
     Why are the "undecided" college students more
attentive to science than those who have chosen a
low or medium prestige occupation, while the "undecided" high school students are less attentive than
those who have chosen either a high or a low to medium prestige occupation? This pattern of relationships suggests that to be "undecided" about one's
occupation in high school may have a very different
meaning than to be "undecided" in college. For college students, being "undecided" about an occupation
may mean that one is still exploring alternatives
that are opening up or that one is putting off the

Table 9.4. Percent Attentive to Science by
Students' Occupational Aspirations and Family
Social Status for Educational Cohorts

Group	Social Status	Occupational Aspiration	Attentive to Science N	Attentive to Science %
HS-No Coll	Lower	Undecided	297	1
		Medium	714	5
		High	46	7
	Higher	Undecided	48	7
		Medium	103	9
		High	10	(0)
HS-Coll Bnd	Lower	Undecided	119	12
		Medium	441	13
		High	138	15
	Higher	Undecided	84	11
		Medium	230	27
		High	85	25
College	Lower	Undecided	90	28
		Medium	530	22
		High	149	38
	Higher	Undecided	89	38
		Medium	403	31
		High	162	37
Total			3738	18

Note: Rounding error due to weighting.

choice of a specific career until one has finished
undergraduate work and is ready to make an informed
decision about what profession or line of business
to enter. Making the decision to pursue a medium or
low prestige occupation after college, on the other
hand, may mean that a student decides to discontinue
his or her formal education after four years of college and starts to turn his or her attention away
from the more abstract public issues to focus more
on the immediate prospects of a particular job.
     To be "undecided" in high school might, by con-
trast, have an exactly opposite meaning. Relative
to the students who have started looking into their

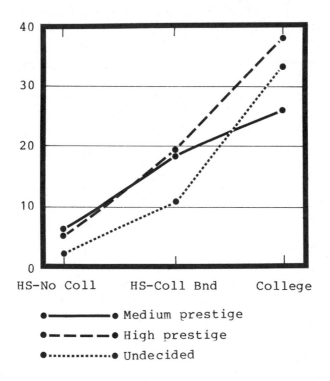

Fig. 9.2.   Percentage attentive to organized science
by educational aspirations and parental occupational
prestige.

futures with some care, as reflected in their having
decided on the occupation they will pursue, the "undecided" high school students may simply be those
who have not yet made up their minds about what they
want out of life and are simply floating along paying little attention to anything as serious as science and technology or other public issues.
     If this characterization of high school and col-
lege "undecideds" is correct, college students who
are undecided about an occupation should have higher
educational aspirations for themselves than students
who have decided on a low or medium prestige occupation. Similarly, high school students who have decided on an occupation should have higher educational aspirations than their undecided colleagues. To
check this possibility, the educational aspirations

of the "undecided" high school and college students
were compared with the educational aspirations of
those in each cohort who had already chosen a
medium or low prestige occupation. Sixty percent
of the high school students who had chosen a medium
or low occupation were aspiring to graduate or professional school, compared to 53 percent of the undecided high school students. However, an almost
identical difference existed among the college students: 63 percent of those who had chosen a medium
or low prestige occupation were aspiring to graduate
or professional school, compared to 54 percent of
those who were undecided. Since there is essentially no difference in this relationship between high
school and college students, there is no evidence
that the difference in attentiveness of "undecideds"
in high school and college can be explained in terms
of differential educational aspirations.

If the answer to why the undecided college stu-
ents are becoming attentive to science is not to be
found in their having higher educational aspirations,

Table 9.5. Model for Data in Table 9.4

Models		LRX^2	df	CMPD
H1	OSX,A	380.09	17	--
H2	OSX,AX,AS,AO	31.50	12	.92
H3	H2 minus AX	240.00	14	--
	Difference due to AX	208.55	2	.55
H4	H2 minus AS	59.44	13	--
	Difference due to AS	27.94	1	.07
H5	H2 minus AO	44.31	14	--
	Difference due to AO	12.81	2	.03
H6	OSX,AX,AS,AO,AOX	13.37	8	--
	Difference due to AOX	18.13	4	.05

A = attentiveness to science
O = occupational aspirations
S = socioeconomic status
X = educational cohort typology

suggesting that they are aiming at higher prestige
occupations eventually, maybe the answer can be found
in the kinds of occupations they are oriented toward
and the kinds of courses they are taking in anticipa-
tion of a general class of occupations. Since there
is no way for us to know what these unreported and
possibly vague occupational aspirations are for the
as yet undecided college students, we have no way of
directly testing this conjecture. In the next chapter, however, we will look at some of the different
kinds of occupations students are expecting to enter
and the effects the focus on specific occupations
has on the development of attentiveness to science.

SEX-ROLE SOCIALIZATION

Research has shown that parents tend to treat sons
and daughters differently with respect to interests
and skills that contribute to the development of
attentiveness to science (Maccoby & Jacklin 1974).
Astin (1971) found in studying parents of sixth,
seventh, and eighth grade students that boys, more
than girls, tend to be encouraged to do well in mathematics and science courses and that boys tend to be
given more encouragement and resources -- in the form
of puzzles, science kits, and the like -- to develop
interests and skills relevant to science. Such differential treatment fosters the development of different habits of intellectual inquiry and persistence
in problem solving, different kinds and levels of
achievement motivation, and different degrees of
independence in thought and action (Rossi 1965).
     When girls are taught by their parents, their
siblings, their peers, and later by their teachers,
that they are not expected to develop the skills
that are prerequisites to the development of interest and knowledge in science, the results are predictable. Girls become less interested than boys in
studying the sciences during elementary and high
school (Omerod & Duckworth 1975); girls are less
likely than boys to be confident of their academic
abilities in mathematics and science ((Oetzel 1961);
girls are less likely than boys to see any relationship between reading or taking courses in mathematics
or science and the world of work they plan for themselves (Hilton & Berglund 1971); and girls are more
likely than boys to become discouraged by low grades
in mathematics and science courses (Bridgeham 1972).

In short, the research shows that girls are quite
likely to avoid science as a topic of interest and,
hence, are unlikely to develop the knowledge necessary to become attentive to science or technology in
their later lives.

The effects of differential sex-role socialization within the family can be seen when we analyze
attentiveness to organized science separately for
men and women (see Tables 9.6 and 9.7). In all but
two of the subpopulations defined by family status,
educational cohort, and occupational aspirations, men
are more likely to be attentive to science than are
women. The conditional contribution of gender differences to the prediction of percent attentive is
very strong and highly reliable (chi-square = 86.43,
df = 1, p<.001). Young women are less likely to be
attentive to organized science than young men of
comparable background, regardless of the social class
in which they were raised, the education they have
obtained or aspire to, or their occupational aspirations. The only exceptions to this general rule are
found among high school men and women from higher
status families who themselves have high occupational aspirations and among college men and women from
lower status families who are undecided about their
occupations. However, the 175 students who fall in
these two groups in the NPAS sample represent only 5
percent of the total population. These results suggest that those factors that operate in the home to
encourage boys to become attentive to science during
their high school and college years do not operate
as strongly for girls (cf. Rossi 1965).

The large and consistent differences in attentiveness between men and women in each of the separate subpopulations described in Table 9.6 should
not, however, be allowed to obscure the fact that
some women do become attentive to science and that
each of the variables of family social status, educational achievement, and occupational aspirations do
have some effect on the development of attentiveness
among women as well as among men. The logit model
of the multivariate relationship among these variables indicates, however, that gender is second only
to educational status and plans in its conditional
association with attentiveness and that these differences in attentiveness between men and women gener-
alize across subpopulations defined by all the other
variables in the model (see Table 9.7).

Table 9.6. Percent Attentive to Science by
Gender, Occupational Aspirations, Family
Status and Educational Cohort

Cohort	SES	Occ/Asp	Gender	Attentive to science N	Attentive to science %
HS-No Coll	Lower	Undecided	Female	157	0
			Male	139	2
		Medium	Female	408	4
			Male	307	7
		High	Female	23	5
			Male	23	8
	Higher	Undecided	Female	34	9
			Male	15	(1)
		Medium	Female	55	0
			Male	48	19
		High	Female	4	(0)
			Male	7	(0)
HS-Coll Bnd	Lower	Undecided	Female	55	10
			Male	63	13
		Medium	Female	246	5
			Male	195	25
		High	Female	65	5
			Male	73	24
	Higher	Undecided	Female	43	3
			Male	41	19
		Medium	Female	115	20
			Male	116	33
		High	Female	36	27
			Male	49	24
College	Lower	Undecided	Female	48	31
			Male	42	25
		Medium	Female	267	15
			Male	262	28
		High	Female	47	28
			Male	103	43
	Higher	Undecided	Female	43	26
			Male	45	49
		Medium	Female	228	24
			Male	175	41
		High	Female	51	23
			Male	111	44
Total				3738	18

RELIGIOUS BELIEFS AND PARTICIPATION

As families tend to socialize children into the "proper" roles for their gender, so families also tend to socialize children to a particular set of religious beliefs and practices. A person's religious beliefs function both as a descriptive model of what the world is like and a prescriptive model for what the world ought to be (McCready & Greeley 1976). When the religious beliefs of a family present images of the world that conflict with the images constructed by scientists and technologists, a tension can develop between paying attention to the tenets of one's religion and accepting other sources of information about the world (Glock & Stark 1965). It is reasonable to expect, then, that the religious outlook that a person is exposed to in the family and which he or she carries into adulthood may play a role in determining whether an individual becomes attentive to organized science.

To investigate this possibility, the NPAS questionnaire included a number of items tapping various aspects of religious beliefs that might conflict with attentiveness to organized science. A factor analysis of the attitude battery (see Table 9.8) identified a cluster of five items that appear to

Table 9.7. Models for Data in Table 9.6

Models		LRX^2	df	CMPD
H1	OSXG,A	495.73	35	--
H2	OSXG,AO,AS,AX,AOX,AG	51.06	25	.90
H3	H2 minus AG	133.97	1	--
	Difference due to AG	86.20	1	.17

A = attentive to science
S = family status
O = occupational aspirations
X = educational cohort typology
G = gender

reflect a general theological perspective:

I believe there is life after death.

I believe there is a God who can and has
personally intervened in the lives of men
and women.

I believe that the Bible is literally true
and we should believe everything it says.

I feel it is important for my children to
believe in God.

Prayer plays an important role in my life.

Agreement with four of the five items is negatively
related to attentiveness to organized science (see
Table 9.9). The strongest negative association is
between attentiveness to science and agreement with
a literal interpretation of the Bible (gamma = -.35).
Acceptance of traditional or fundamentalist religious images of the world appears to deflect one
away from becoming attentive to science.
    The consistent pattern of association suggests
that different aspects of religious belief are in-
compatible with attentiveness to science for different people, but that no single belief necessarily
reduces the likelihood that one will become attentive. Under these circumstances, it is reasonable
to combine responses to the individual items into an
index of General Religious Belief. To construct this
index, the sample was divided into those people who
answered "agree" or "strongly agree" to at least
four of the belief items versus those people who
answered "uncertain," "disagree," or "strongly disagree" to two or more of these items. The first
group was classified as "more religious" on the
Index of General Religious Belief, while the latter
was classified as "less religious." Fourteen percent of the "more religious" were attentive to organized science, compared to 23 percent among the
"less religious." Since most of even the "less religious" students in the NPAS sample indicated some
agreement with certain of the religious belief statements, the difference in percent attentive between
these two groups suggests the existence of conflicts,
or tensions, between religious and scientific out-
looks that apparently contribute to a lack of inter-

Table 9.8.   Factor Analysis of Religious Beliefs

		Factor loadings	2 h
a.	Belief in life after death	.55	.30
b.	There is a God	.90	.81
c.	The Bible is literally true	.71	.50
d.	My children should believe in God	.92	.85
e.	Prayer is important in my life	.84	.71
Proportion of common variation accounted for:		.70	

est and knowledge of science among the more religious.

Besides their religious beliefs, the students
were also asked to indicate the frequency of their
attendance at religious services. The relationship
between frequency of attendance and attentiveness to
organized science was extremely weak and slightly
negative. This result suggests that it is indeed
religious belief that is primarily responsible for
the differences in attentiveness -- not frequency of
actual participation.

A logit model was constructed to examine the
association between religious belief (as measured by
the index) and attentiveness to organized science in
the context of denomination and frequency of attendance (see Table 9.10). The data indicate that there
are no substantial interactions among the dimensions
of religiosity in predicting attentiveness to science
(see Table 9.11). There is no evidence to support
the idea that the association between religious belief and attentiveness to science is conditional on
either the kind of religious services one is exposed
to or the frequency of one's attendance. Students
who report strong religious beliefs are more likely
to attend religious services regularly, and those
who report affiliations with either Protestant or

Table 9.9. Percent Attentive by Religious Affiliation, Attendance at Religious Services, and Reported Religious Beliefs

Religious beliefs	N	Strongly Agree	Agree	Un-certain	Dis-agree	Strongly Disagree	Gamma
a. Life after death	3785	17 (1481)	17 (1085)	18 (749)	14 (324)	22 (236)	.03
b. There is a God	3860	14 (1714)	16 (1274)	23 (472)	27 (272)	30 (128)	-.22
c. The Bible is literally true	3875	10 (1251)	13 (967)	17 (499)	23 (897)	31 (460)	-.35
d. My children should believe in God	3852	14 (1673)	18 (1384)	20 (346)	20 (312)	34 (137)	-.18
e. Prayer is important in my life	3825	15 (1019)	13 (1223)	18 (360)	20 (870)	30 (353)	-.17

"other" non-Catholic and non-Jewish denominations are
more likely to agree with the religious belief statements above. However, attending a Protestant
church does not appear to make the association between these religious beliefs and attentiveness any
stronger or weaker than attending a Catholic church
or a Jewish synagogue, or not attending religious
services at all.
    To understand the relationship of religious be-
liefs with attentiveness to science, it is necessary
to consider which kinds of families are most likely

Table 9.10. Percent Attentive to Science by
Religious Denomination, Attendance, and Beliefs

Denomination	Attendance	Religious Beliefs	Attentive to Science	
			N	%
Protestant	Rare	Less	328	21
		More	296	13
	Often	Less	120	13
		More	594	14
Catholic	Rare	Less	284	19
		More	180	17
	Often	Less	178	26
		More	438	19
Jew	Rare	Less	106	41
		More	17	(3)
	Often	Less	9	(1)
		More	8	(2)
Other	Rare	Less	150	19
		More	126	5
	Often	Less	63	17
		More	360	11
None	Rare	Less	329	24
		More	55	13
	Often	Less	14	(1)
		More	23	4
Total			3678	18

Table 9.11. Models for Data in Table 9.10

Models		LRX^2	df	CMPD
H1	BCD,A	100.79	19	--
H2	BCD,AB,AC,AD	24.54	13	.76
H3	H2 minus AB	44.97	14	--
	Difference due to AB	20.43	1	.20
H4	H2 minus AC	24.68	14	--
	Difference due to AC	.15	1	.00
H5	H2 minus AD	60.08	17	--
	Difference due to AD	35.54	4	.35

A = attentiveness to science
B = religious beliefs
C = attendance at religious services
D = denomination

to produce children with strong religious beliefs,
and how much these religious beliefs affect attentiveness over and above the influence of these other
family characteristics. Students from lower status
families are more likely to express agreement with
the five religious beliefs than are students from
higher status families, while college students are
less likely to express agreement than are high school
students (see Table 9.12). The logit model that fits
these data, however, indicates that a student's religious beliefs have little residual association
with attentiveness to science once the student's
social status, educational status, and educational
and occupational aspirations are taken into account
(see Table 9.13). The fact that students with strong
religious beliefs are underrepresented among college
students and among students from higher status families appears to account for most of the association
we have seen between religious belief and the tendency not to become attentive to organized science.

Table 9.12. Percent Attentive to Science by
Religious Beliefs, Occupational Aspirations,
Family Status, and Educational Group

Group	SES	Occ/Asp	Religious Beliefs	Attentive to Science N	Attentive to Science %
HS-No Coll	Lower	Undecided	Less	107	1
			More	173	2
		Med/Low	Less	221	5
			More	425	5
		High	Less	11	(1)
			More	29	0
	Higher	Undecided	Less	15	(3)
			More	30	3
		Med/Low	Less	35	11
			More	63	8
		High	Less	5	(0)
			More	5	(0)
HS-Coll Bnd	Lower	Undecided	Less	44	3
			More	70	13
		Med/Low	Less	147	17
			More	279	12
		High	Less	52	12
			More	82	18
	Higher	Undecided	Less	42	12
			More	39	10
		Med/Low	Less	97	37
			More	130	19
		High	Less	25	32
			More	52	19
College	Lower	Undecided	Less	56	38
			More	35	14
		Med/Low	Less	253	26
			More	267	19
		High	Less	85	46
			More	61	26
	Higher	Undecided	Less	65	42
			More	23	26
		Med/Low	Less	223	33
			More	171	31
		High	Less	89	28
			More	70	50
Total				3603	18

FAMILY POLITICIZATION

In the introduction to this chapter, it was suggested that the family influences the development of
attentiveness to science in three ways -- by providing access to resources, by providing models of attentiveness in other people, and by direct transmission of interests and knowledge from parents to
children.
    The present section focuses on the third possi-
bility -- that some students become attentive to
science, in part, through the direct transmission of
interest and information from other members of their
families. The measure of direct transmission employed is the reported frequency of discussion of
public issues within the family. Lane (1959),
Wrightsman (1970), Tolley (1973), Jennings & Niemi
(1974), and Beck (1977) all agree that frequency of
discussion of public issues in the home is a key variable that links the interests and knowledge of parents with the political views of their children.
Without discussion, it is unlikely that children will
recognize the attitudes their parents hold or perceive which issues their parents care about. And,
as Tedin (1974) has demonstrated, unless parents are
interested enough themselves in public issues to
discuss them frequently, their effectiveness as

Table 9.13.   Models for Data in Table 9.12

Models	LRX^2	df	CMPD
H1 OSXR,A	396.32	35	--
H2 OSXR,AO,AS,AX,AOX,AR	47.18	25	.88
H3 H2 minus AR	57.56	26	--
Difference due to AR	10.37	1	.03

A = attentiveness to science
O = occupational aspirations
X = educational cohort typology
S = family status
R = religious belief index

transmitters of information on these issues will
be diminished.

As a measure of frequency of family discussions,
respondents were asked to report the number of times
"so far this school year" they had discussed each of
four issues -- foreign policy, economic policy, science, and civil rights -- with other members of
their families. Respondents indicating that two or
more of these issues were discussed three or more
times within the family were classified as having a
high level of family politicization. Using this
criterion, 41 percent of the students were classified as high on family politicization, while 59 percent were classified as low on family politicization.

Work by Chaffee and his associates (1977, 1971,
1966) indicates that both frequency of family discussions and the role the children play in these discussions vary with the social status of the family. It
will be useful, therefore, to examine the association between the frequency of family discussion and
attentiveness both in the aggregate and within the
social stratum of each family.

Controlling for family social status and the
students' educational aspirations, frequent family
discussions made a large contribution to the likelihood that a student will become attentive to organized science during high school or college (see
Table 9.14). There is an average of 16 percent difference in the proportion attentive comparing students with high and low levels of family politicization, and this difference does not vary significantly among students from different family backgrounds
or with different educational or occupational aspirations.

Log-linear analysis of this relationship con-
firms the reliability of this finding and indicates
that approximately 15 percent of the mutual dependence in the multivariate model is accounted for by
the direct effect of family politicization on attentiveness (see Table 9.15). Regardless of the other
influences to which these students have been exposed,
family discussions remain pivotal in determining the
likelihood that the student will become attentive to
science during his or her high school or college
years.

Table 9.14. Percent Attentive to Science by
Frequency of Family Discussions, Occupational
Aspirations, Family Status, and Educational Cohort

Group	Family Status	Occ/Asp	Family Discussion	Attentive to Science N	%
Hs-No Coll	Lower	Undecided	Infrequent	238	1
			Frequent	59	3
		Medium	Infrequent	520	3
			Frequent	195	12
		High	Infrequent	37	5
			Frequent	9	(1)
	Higher	Undecided	Infrequent	30	10
			Frequent	19	5
		Medium	Infrequent	64	8
			Frequent	39	10
		High	Infrequent	6	(0)
			Frequent	4	(0)
Hs-Coll Bnd	Lower	Undecided	Infrequent	72	6
			Frequent	48	21
		Medium	Infrequent	234	7
			Frequent	207	20
		High	Infrequent	70	11
			Frequent	68	19
	Higher	Undecided	Infrequent	49	2
			Frequent	35	23
		Medium	Infrequent	118	18
			Frequent	113	35
		High	Infrequent	34	18
			Frequent	51	29
College	Lower	Undecided	Infrequent	37	22
			Frequent	43	40
		Medium	Infrequent	277	17
			Frequent	253	27
		High	Infrequent	74	36
			Frequent	75	40
	Higher	Undecided	Infrequent	27	37
			Frequent	52	46
		Medium	Infrequent	215	26
			Frequent	183	38
		High	Infrequent	69	35
			Frequent	94	39
Total				3742	18

A PROCESS MODEL OF FAMILY EFFECTS

The preceding section has described how the proportion of students attentive to science varies with a
number of family and individual characteristics. In
each analysis, the relationships were examined in a
multivariate context, controlling for the association among the various family effects to remove any
confounded effects from the predictive models. In
short, the analysis has focused on determining which
family characteristics contribute to the development
of attentiveness to science, not on how these individual characteristics affect each other and mediate
each other's influences.
     In this final section, the focus of the analy-
sis will turn to specification of the relationships
among family characteristics, utilizing a causal
path analysis. To combine all the family predictors
into a single model, however, requires that we identify the assumptions being made about the causal order in which these predictors come into play. For
the present analysis, the following assumptions are
made:

1. The socioeconomic status of the family,
    based on parental educational level and
    occupational prestige, exists prior to
    the development of the student's educa-

Table 9.15. Models for Data in Table 9.14

Models	LRX^2	df	CMPD
H1 OSXF,A	459.91	35	--
H2 OSXF,AO,AS,AX,AOX,AF	37.14	25	.93
H3 H2 minus AF	104.46	26	--
Difference due to AF	67.32	1	.15

A = attentiveness to science
O = occupational aspirations
S = family status
X = educational cohort typology
F = frequency of family discussions

tional and occupational aspirations. Family status may influence the student's educational and occupational aspirations, but
the student's educational and occupational
aspirations do not influence the family's
social status.

2. Sex-role socialization practices indexed
by the student's gender are a characteristic of the family and have their major
impact prior to the development of the student's educational and occupational aspirations.

3. The student's religious beliefs may be influenced by the social status of the family and the sex-role socialization to
which the student is exposed, but the student's religious beliefs have no reciprocal effect on either the family's status
or the family's socialization practices.

4. Frequency of family discussions of public
issues may be influenced by the family's
social status and whether a particular
member of the family participates in those
discussions may be influenced by the family's sex-role socialization practices,
but frequency of family discussions and
the participation of individual members
of the family have no influence over the
family's status or socialization practices.

5. The causal relations among the student's
educational and occupational aspirations
and achievement are probably reciprocal
and correlated with the frequency of family discussions and the religious beliefs
of the student.

6. Becoming attentive to science and technology is an outcome of the variables discussed above and is not reciprocally related to any of them.

Assumption six calls for some clarification.   In
looking for the characteristics of families and
individuals that "influence" the development of
attentiveness to science and technology, the analysis
has concentrated on factors that are likely to exist
temporally prior to the student's becoming interested in and informed about public issues or which
can meaningfully be thought of as representing
broader, more basic characteristics of the individual out of which interest and knowledge might devel-

op. The assumption that attentiveness to science is
a causal result of these factors is simply an expli-
cation of the assumptions underlying much of the
previous analysis. This assumption is not an asser-
tion that family discussions, religious beliefs,
educational achievements, or occupational aspirations cannot be influenced by the level of attentiveness to science or other issue areas, but rather a
judgment that the preponderance of the influence
moves in the other direction. Only a longitudinal
study focusing specifically on the order of development of these variables could measure how much
influence each variable has on the other. For the
purposes of speculating about these causal processes, however, it will be useful to assume a
one-way causal process so that all of the residual
association among these variables can be attributed
to the causal impact of the more general characteristics on the more specific.

Looking first at the socioeconomic status of
the family, the NPAS data indicate that the strongest
effect of social class is the transmission of educational aspirations, as measured by the educational
plans typology. The standardized value (S.V.) for
the relationship between higher social class and
college attendance is 8.0, which is one of the strongest relationships in the entire model (see Fig. 9.3).
Family social status also affects the transmission
of occupational aspirations, which are associated
with educational aspirations. Children of higher
status families are significantly less likely than
their lower status counterparts to select a middle-
to-lower prestige occupation (S.V. = -2.3), and the
decision to pursue a higher status occupation is
significantly associated with college attendance
(S.V. = 6.9). College attendance, in turn, is
strongly and positively associated with attentiveness to organized science (S.V. = 6.4).

In addition to the above paths, higher social
status has a significant and positive direct effect
on attentiveness (S.V. = 2.8), indicating that other
facets of social class than those expressed in the
transmission of educational and occupational aspirations are operative. What these factors are -- and
whether they reflect differences in the resources
higher status families can provide their children,
differences in the levels of attentiveness modeled
in higher status families, or differences in the
processes by which higher status families teach the
interests and skills underlying attentiveness --

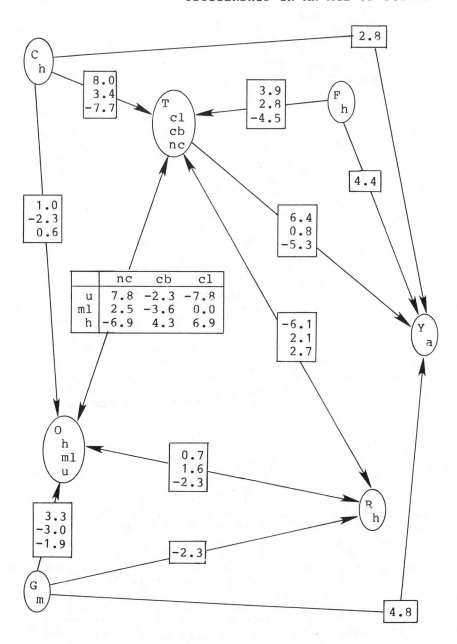

Fig. 9.3. A path model of the relationship between attentiveness to organized science and selected family-related variables.

will have to be the focus of future research on the
development of attentiveness to science.

The second major influence family has on the
development of attentiveness appears to operate
through the differential socialization practices to
which girls and boys are exposed. The gender of the
respondent has a significant and direct effect on
attentiveness (S.V. = 4.8), controlling for all
other predictor variables in the model. Interestingly, this large direct "effect" of gender is not due
to a lack of correlation between gender and the
other predictors, but is due to the fact that the
variables with which gender is directly associated
do not themselves have any direct "effects" on attentiveness.

For example, young men are significantly more
likely to choose higher status occupations than are

Legend for Fig. 9.3.

Y = attentiveness to organized science
 a   attentive

C = index of family socio-economic status
 h   upper half

G = gender
 m   male

O = occupational aspiration
 h   high
 ml   medium-low
 u   undecided

T = educational plans typology
 cl   college
 cb   HS-coll bnd
 nc   HS-no coll

F = level of family politicization
 h   high

R = religious belief index
 h   high

young women (S.V. = 3.3), and persons pursuing higher status occupations are significantly more likely
to attend college (S.V. = 6.9). However, since
higher status occupations do not have a signiificant
direct effect on attentiveness, the relationship
between gender and occupational aspirations at best
represents only part of a longer indirect path of
influence through educational aspirations.

Similarly, young women score significantly
higher on the index of religiosity than young men
(S.V. = -2.3), and persons scoring high on this
index are significantly less likely to attend college (S.V. = -6.1). But there is no direct effect
of religious beliefs on attentiveness. Again,
whatever "effects" gender related differences in
religious beliefs have on attentiveness must be
mediated indirectly through influences on educational
aspirations.

The third major influence families have on the
development of attentiveness to science is through
discussions of public issues within the home. The
level of family politicization is positively and
significantly associated with attentiveness to organized science (S.V. = 4.4), controlling for all of
the other predictor or variables in the model. In
addition, the frequency of family discussions is
strongly correlated with the student's educational
status and aspirations. The NPAS data indicate that
college-bound high school students and college students are significantly more likely to engage in
family political discussions than are non-college-
bound students (S.V.'s = 2.8, 3.9, -4.5, respectively). The data do not, however, indicate whether
these discussions are primarily parent-generated or
child-generated, which would suggest the causal
order of these socializing influences within the
family. It is important to note that, controlling
for educational status and aspirations, the level of
family politicization is not significantly related
to social class, gender, or religiosity. Families
that talk about science and technology issues are
more likely to foster attentiveness to these issues
in their children regardless of the social class of
the family, the sex of the child, or the religious
socialization that occurs within the family.

The educational-plans typology is the final
variable in the model that needs to be discussed.
Throughout this analysis, major differences in attentiveness among the three educational plans cohorts
has appeared. The model diagrammed in Fig. 9.3 re-

iterates the central role of this variable. The
model illustrates why educational status and aspirations are so important and suggests how differences
in status and aspirations impact on the development
of attentiveness.
     Students in high school and college with differ-
ent educational aspirations tend to differ on all
the other predictive factors except gender. Students in college and those planning to go to college
disproportionately come from the higher status families. These students also are more likely to be
aspiring toward a high-status occupation and more
likely to be talking to their families about public
issues. In addition, college students are significantly less likely to hold strong religious beliefs
that appear to conflict with the tenets of science
and with developing an interest in and attentiveness
to scientific issues. Taken together, all these
factors appear to contribute to the fact that college students are significantly more likely to be
attentive to science than are high school students
and that among high school students, those who are
planning to attend college are more attentive to
science than are those who have no college plans.
This finding suggests that it is important to investigate the other factors associated with aspirations
and exposure to higher education to better understand the processes by which young people become
attentive to organized science.

10 School and Peers

Beyond the family, the literature suggests that schools and student peers are the next most important agents of socialization. It is the purpose of this chapter to explore how students' experiences in school influence the development of attentiveness to organized science.

The most obvious school attribute to consider is the academic achievement of the student. Studies have shown that academic achievement is positively related to the development of a wide variety of interests and knowledge outside the classroom (Allen 1959; Belt 1959; Bixler 1958). It is reasonable to suggest, therefore, that students who exhibit a high level of academic competence will be more likely to be attentive to organized science than students who are less successful academically. In the analysis below, a measure of academic achievement -- class rank -- will be examined as a predictor of attentiveness.

The heart of the school experience is formal instruction. It is anticipated that the level of exposure to science instruction should be positively associated with development of each of the three components of attentiveness -- interest, knowledge, and information acquisition. Myers (1967), however, found no relationship between science background and attitudes toward science among college students. The number of science courses taken and the number of disciplines covered will be studied in relationship to attentiveness.

Another way that school programs may influence the development of attentiveness is through the dis-

cussion of issues. Many teachers and professors
incorporate discussions of various political and
social issues into their class activities. Out-of-
class discussions among peers may further contribute
to the development of attentiveness. Discussion is
an excellent way for students to share interests,
acquire knowledge not available through standard
sources, and confront opinions different from those
received from family sources. Participation in issue discussions is a major source of politicization
(Jennings and Niemi 1974) and should be expected to
contribute to the development of attentiveness to
science as well as to other public issues.

During the high school and college years, many
students will begin to make decisions about life
goals and future plans. While these decisions will
be influenced in part by family values, the literature on career choice suggests that the schools play
a substantial role in this process (Davis 1965). In
chapter nine, it was suggested that the social status experienced in the family influences the general
prestige level of a student's occupational aspirations. If schools broaden the perspectives of students, it is likely that experiences in school
influence the choice of specific occupations and
professions (Davis 1965) and these choices, in turn,
may influence whether or not a student begins to develop an attentiveness to science commensurate with
his or her occupational choice.

THE BASE FAMILY EFFECTS MODEL

In chapter nine, three variables were identified
that exhibited a substantial effect on the development of attentiveness to science and technology:
educational status and plans, gender, and the level
of family politicization. These three variables constitute the components for a base model of family
effects.

The logit model for this relationship indicates
that the main -- or direct -- effects of educational
plans, gender, and family politicization account for
96 percent of the total mutual dependence in the model, which is an extraordinarily good fit for the multivariate table (see Table 10.1). The strongest
main effect comes from educational plans, which
accounts for about 48 percent of the total mutual
dependence in the model. Gender is a strong main

effect, accounting for about 19 percent of the mutual dependence in the model, while family politicization explains about 15 percent. (1)

ACADEMIC ACHIEVEMENT AND SCIENCE COURSEWORK

Formal instruction is the heart of the school experience and the traditional focus of studies of school effects. In the 1978 NPAS, the instructional aspect of schooling was tapped by two measures -- general academic achievement and scope of science course work.

It is interesting to note that academic achievement and the scope of science study are not significantly correlated in the multivariate model. Since the two variables are significantly correlated in the zero-order relationship, it is necessary to conclude that the bivariate relationship is spurious, apparently reflecting differences in educational plans, gender, and family politicization. This finding runs counter to the long-held view that only the best and brightest students attempt science courses.

Table 10.1. Chi-Square Values for the Base Family Effects Model

Model	Fitted Parameters	df	LRX^2	CMPD
H1	FGX,A	11	456.10	--
H2	FGX,AF,AG,AX	7	17.38	.96
H3	H2 minus AF	8	93.52	--
	difference due to AF	1	76.14	.15
H4	H2 minus AG	8	112.98	--
	difference due to AG	1	95.60	.19
H5	H2 minus AX	9	252.01	--
	difference due to AX	2	234.63	.48

A = attentiveness to organized science
F = family politicization
G = gender
X = educational status and plans

General academic achievement was measured by
the student's self-report of class standing. Respondents were asked to indicate whether they were in
the top 5 percent,. the next 5 percent, the upper
quarter, the upper half, or the lower half of their
school or college. The measurement of academic
achievement between and among different institutions
is a difficult task. Lacking any national measures
analogous to the New York Regency Examination, the
two best alternatives were the student's class standing or his grade-point average (GPA). Both measures
were collected on the 1978 NPAS; subsequent analysis
indicated that class standing provided substantially
more variation, and it was adopted for analytic purposes.

An analysis of the relationship of class rank
to each of the three components of attentiveness and
to the composite attentiveness measure indicates
that class rank is primarily associated with science
and technology knowledge (see Table 10.2). Using
the ordinal correlation coefficient gamma, the NPAS
data show a strong positive relationship between
knowledge and class rank for all educational cohorts
in the study. Only among college-bound high school
students is class rank a useful predictor of interest in and attentiveness to organized science.

When class rank is examined in the context of
family effects, the logit model indicates that the
level of academic achievement (2) makes a small marginal contribution to the base model, accounting for
only 2 percent of the total mutual dependence in
the model (see Table 10.3). No higher-order or interaction terms involving academic achievement were
significant at the 0.01 level. In short, general
academic achievement displays a relatively weak association with attentiveness to organized science,
especially when viewed in the context of the family
effects which presumably preceded the schooling
experience.

A second major facet of school instruction is
the exposure to science courses. In both high
school and college, students have some latitude in
selecting courses, and these selections may either
reflect emergent interests or stimulate new interests. It is reasonable to expect that the number
and scope of science courses would be related to attentiveness to organized science.

Students indicated on the NPAS questionnaire
how many semesters of study they had completed in
biology, chemistry, physics/physical science, and

Table 10.2. The Relationship Between Academic
Achievement and Science Coursework and the
Components of Attentiveness
to Organized Science

Group	Academic Achievement	Science Course Exposure
HS-No Coll		
Interest	.09	.04
Knowledge	.25*	.25*
Consumption	.09	.04
APST	.16	.10
HS-Coll Bnd		
Interest	.05*	.25*
Knowledge	.32*	.31*
Consumption	.04	.21*
APST	.23*	.26*
College		
Interest	.05	.20*
Knowledge	.30*	.38*
Consumption	.07	.05
APST	.10	.14*

* Significant at p < .01. All others not
significant at p < .05.

natural science. Natural science was employed as a
label for a category that included science courses
such as earth science, environmental science, and
other more recent creations outside the standard
areas of biology, chemistry, and physics. Preliminary analyses of these data indicated that it would
be useful to group the number of courses into years
of study for each area. Students were classified as
having had zero, one, or more than one year of study
in each natural science area.

Initially each of the four science areas included in the NPAS study was analyzed separately.
Since many high schools offer only one year of physics, chemistry, and biology, the basic division became some and none. This dichotomy was not a useful

predictor of attentiveness in any of the four science subject areas. As an alternative, the science
course variable was reconstructed to reflect the number of scientific disciplines studied. This scope-
of-science-exposure measure proved to be positively
associated with attentiveness among both high school
and college students.

An analysis of the relationship between the number of scientific disciplines studied and each of
the attentiveness components reveals a number of interactions with the student's educational status and
plans (see Table 10.2). For non-college-bound high
school students, a wide exposure to scientific disciplines is positively associated with a higher
level of cognitive information about science, but unrelated to either interest or information acquisition.
Not surprisingly, the ordinal correlation between
the number of science areas studied and the composite attentiveness measure is not significant for the
non-college-bound population.

For college-bound high school students, however,
the scope of scientific course exposure is positively and significantly associated with each of the components of attentiveness and with the composite
index (see Table 10.2). These results are consistent with the patterns of interest, knowledge, and
information discussed in chapters five and six.

For college students, who have the greatest latitude in elective course selection, the number of

Table 10.3. Chi-Square Values for
Academic Achievement

Model	Fitted Parameters	df	LRX^2	CMPD
H1	TFGX,A	23	499.87	--
H2	TFGX,AT,AF,AG,AX	18	33.89	.93
H3	H2 minus AT	19	42.73	--
	difference due to AT	1	8.84	.02

A = attentiveness to organized science
F = family politicization
G = gender
X = educational status and plans
T = academic achievement (class rank)

scientific disciplines studied is positively and significantly associated with the levels of interest
and knowledge and with the composite index, but not
with the level of information acquisition (see Table
10.2). The absence of any association with information acquisition appears to have the effect of
substantially reducing the correlation with the
composite index.

When the scope of scientific study (3) is examined in the context of the family effects model, the
logit model indicates that its marginal contribution
is low, accounting for only about 2 percent of the
total mutual dependence (see Table 10.4). This is a
particularly surprising and disappointing result, in
view of the substantial level of resources devoted
to curriculum development and course improvement in
recent decades.

Reviewing the effects of the formal instructional program of the school, several conclusions emerge.
First, both general academic achievement and the
scope of scientific disciplines studied appear to relate to students level of cognitive science knowledge, reflecting the long-standing view of the
school as the transmitter of the society's stock of
acquired information. Among high school students,
this relationship holds regardless of expectations
about college.

Table 10.4. Chi-Square Values for Number of
Science Areas Studied

Model	Fitted Parameters	df	LRX^2	CMPD
H1	TFGX,A	23	512.34	--
H2	TFGX,AT,AF,AG,AX	18	32.98	.94
H3	H2 minus AT	19	44.68	--
	difference due to AT		11.70	.02

A = attentiveness to organized science
F = family politicization
G = gender
X = educational status and plans
T = number of science disciplines studied

Second, the level of general academic achievement is unrelated to interest in organized science
in all educational cohorts, suggesting that "bright
ness" or school success does not lead automatically
to a curiosity about or interest in science or technology issues.

Third, neither academic achievement nor scientific breadth is associated with a high level of
information acquisition (except for science courses
among the college-bound high school cohort), indicating that this type of behavior is apparently not
stimulated by the formal instructional program of
schools or colleges.

Finally, the general pattern suggests that
school instructional programs are successful in
transmitting cognitive science information, but that
these programs either have not been successful in
stimulating a science and technology policy interest,
or have not attempted to do so.

POLITICIZATION

Politicization refers to the processes by which
persons develop their attitudes and behavioral dispositions towards politics. The process of politicization is life-long and reflects a large number
of influences, ranging from inherited characteristics like gender to family or peer influences or the
circumstances of the times -- events like the great
depression or a world war. While the process of
identifying the relative influence of any one source
is difficult, the use of multivariate models with
large data sets like the NPAS make it possible to
look for patterns of associations within common
groups that may be indicative of paths or sources of
politicization.

In this section, the effects of the discussion
of political and public policy issues in the school
classroom and among peers outside the classroom will
be examined. The role of the school in political
socialization has been the subject of extensive
study by both political scientists and by educators,
and considerable disagreement still exists over the
relative influence of the school in the overall process. Hess and Torney (1967) assert that the school
is a major force in the development of political
attitudes and that the social studies curriculm provides a major input for policy views other than pol-

itical partisanship. On the other hand, studies
targeted on the influence of civics courses in high
school (Langton and Jennings 1968) have found little
or no change in students attitudes toward the political system. Langton and Jennings (1968) conclude
that:

> there is a lack of evidence that the civ-
> ics curriculum has a significant influence
> on the political orientations of the great
> majority of American high school students.
> Morever, those who are college bound al-
> ready have different political orienta-
> tions than those who do not plan to attend
> college. These two conclusions suggest
> that an important part of the difference
> in political orientations between those
> from different levels of education, which
> is frequently cited in the literature and
> is usually acscribed to the 'educational
> process', may actually represent a serious
> confounding of the effect of selection with
> that of political socialization (p.866).

In addition to the teacher-directed and organized classroom discussions of political and public
policy issues, the discussion of these issues among
peers outside the classroom and away from the family
setting may provide a more voluntary and open forum.
The discussion of political issues among peers is an
important source of politicization, although one little studied in the literature. Jennings and Niemi
(1974) point to the special place of peer influence
during the adolescent years:

> The peer group is particularly important for
> children in modern societies, where most
> nuclear families are small and live phys-
> ically apart from other relatives. Chil-
> dren in such situations grow up having
> primarily nonfamilial friends who do not
> necessarily share common values and loyal-
> ties with them as would siblings and cous-
> ins. The importance of peer relations is
> emphasized at adolesence when the individ-
> ual is of an age to begin abandoning de-
> pendency on the family of origin, but
> discouraged by the cultural patterns of
> modern societies from marrying and assum-
> ing adult roles. In these transitional

years the peer culture provides a supple-
mentary point of reference for the
individual who is seeking self-definition
and a set of independent values (p.230).

Given the analysis procedures utilized in preceding sections, it is possible to examine the relative influence of school politicization and peer
politicization within the context of the pre-exist-
ing parameters of family politicization, gender, and
educational plans and status. For the purpose of
this analysis, both school and peer politicization
were defined by the number of issues that the respondent discussed during the school year in a classroom setting or with peers outside the classroom.
This approach is parallel to the measure of family
politicization discussed in the preceding chapter.
Like family politicization, both school and peer
politicization were dichotomized into a high group
that discussed two or more of the four issue areas
at least three times and a residual low group.
     Looking first at school politicization, the
NPAS data indicate that classroom discussions have
little impact on the development of attentiveness to
organized science (see Table 10.5). When the effects of educational plans, gender, family and peer
politicization are held constant, school politicization adds only about 1 percent of additional
explanatory power to the log-linear model. This result tends to follow the findings of Langton and
Jennings (1968) and suggests that most of the differences in attentiveness to organized science must be
attributed to factors other than the level of school
classroom discussion of political issues.
     Turning to peer politicization, the NPAS data
indicate that the frequency of peer discussions of
political issues has a positive and significant impact on the development of attentiveness to science
and technology issues (see Table 10.5). When peer
politicization is added to the previous model, the
predictive ability of the model increases by about 6
percentage points, a significant increase in the per-
formance of the model. It should be noted that this
increase in predictive power is additive to the minimal increment provided by school politicization.
This finding would indicate that one of the major
sources of influence outside the family during adolescence is the peer group.
     The NPAS data also reveal that the model dis-
cussed above includes a significant interaction be-

Table 10.5. Logit Models Pertaining to School
and Peer Politicization

Model	Fitted Parameters	df	LRX^2	CMPD
H1	PSFGX,A	47	573.05	--
H2	PSFGX,AF,AG,AX	43	107.19	.81
H3	H2 plus AS	42	102.68	--
	difference due to AS	1	4.51	.01
H4	H3 plus AP	41	67.06	--
	difference due to AP	1	35.62	.06
H5	H4 plus APF	40	53.97	--
	difference due to APF	1	13.09	.02

P = peer politicization
S = school politicization
F = family politicization
G = gender
X = educational plans and status
A = attentiveness to organized science

tween family politicization and peer politicization
(see Table 10.5). An examination of the three-way
table reveals that when the level of family politicization is high, the level of peer political discussion makes little difference in the level of
attentiveness to science and technology issues, but
that in the absence of a high level of family political discussion, peer influence is strongly associated with the attentiveness to organized science.
While the interaction term accounts for only an additional 2 percent of predictive power for the model,
the specification of the circumstances in which family and peer influences are effective is an important contribution.
    In summary, the NPAS data provide an important
opportunity to examine the relative role of school
and peer politicization within the context of the
family effects model. The results indicate that
school politicization has no significant effect on
the development of attentiveness to organized science, but that peer politicization is a positive

factor in the development of attentiveness to science and technology issues. The data specify that
the major effect of peer influence is to be found
when the level of family political discussion is low.

LIFE GOALS AND OCCUPATIONAL PREFERENCE

During the high school and college years, young
people make a number of important decisions about
the major goals they wish to pursue in their lives.
They may decide on both a general career area and
make a tentative selection of a specific occupation
or profession. They may choose a spouse and begin a
family, or they may make a decision to postpone or
not pursue family goals. The process of thinking
about these important decisions during this period
may contribute significantly to the young person's
selection of fields of political interest or issue
specialization.
    The NPAS data show that young persons expecting
to pursue a career as a scientific researcher have
the highest level of attentiveness to organized science -- 44 percent. Approximately 35 percent of prospective physicians and 30 percent of prospective
engineers are attentive to science and technology
issues. Students anticipating public service careers were also significantly more likely to be attentive to organized science than most other
occupational classifications.
    To explore the role of the formulation of life
goals and career plans on the development of attentiveness, the following analysis will examine the
relative importance of family and career objectives
to respondents and will study the pattern of attentiveness among the various prospective occupational
groupings. Using the multivariate framework utilized above, it will be possible to explore the marginal contribution of life goals and career plans
within the context of the family influence parameters discussed in previous sections.

Life Goals

One of the most important decisions facing a young
person is the relative importance of family and
career aspirations. In the 1978 NPAS, each respon-

dent was asked to assess the relative importance
of eleven life goals; a factor analysis of the resulting data found three factors, which were labeled "interest in family," "interest in career,"
and "interest in politics." The political interest variable will be discussed in chapter twelve,
but the family interest and career interest dimensions are important to the present analysis.
     Career interest was measured by a set of five
items that formed a single factor in the factor
analysis noted above. Each NPAS respondent was asked
to evaluate whether, looking into the future, each
of the following would be essential, very important,
somewhat important, or not important as a life goal:

     Becoming accomplished in my career field.

     Becoming well off financially.

     Becoming an authority in my field.

     Having administrative responsibility over
     others' work.

     Obtaining recognition from colleagues for
     contributions to my special field.

Respondents who indicated that three or more of
these items were either essential or very important
were classified as high on the index of career interest and the residual group was classified as low on
on the dichotomized measure. Approximately 55 percent of the NPAS respondents scored high on this
index.
     To assess the marginal contribution of career
interest to the development of attentiveness to organized science, a multivariate log-linear model was
used that holds constant the relationship among educational plans, gender, family politicization, occupational preference, family interest, and career
interest, and assesses the relationship between career interest and attentiveness to organized science. The NPAS data indicate that the marginal
contribution of career interest is minimal, accounting for only 1 percent of the predictive power of
the model (see Table 10.6).
     Family interest was measured by a single item.
Each respondent was asked to indicate whether "rais-
ing a family" was essential, very important, some-

Table 10.6. Logit Models Relating to Life Goals
           and Occupational Preferences

Model	Fitted Parameters	df	LRX^2	CMPD
H1	OSCFGX,A	95	610.50	--
H2	OSCFGX,AF,AG,AX	91	162.17	.73
H3	H2 plus AC	90	156.78	--
	difference due to AC	1	5.39	.01
H4	H3 plus AS	89	151.88	--
	difference due to AS	1	4.90	.01
H5	H4 plus AO	88	110.19	--
	difference due to AO	1	41.69	.07

C = career interest
S = family interest
O = occupational preference
F = family politicization
G = gender
X = educational plans and status
A = attentiveness to organized science

what important, or not at all important to his or
her life goals. Students indicating that a family
was either essential or very important were coded as
high on family interest, and others were classified
as low on the variable. Approximately 60 percent of
the NPAS respondents scored high on family interest.
    When the level of interest in a family is added
to the model discussed above, the NPAS data indicate
that the marginal contribution of family interest is
minimal, accounting for only 1 percent of the total
mutual dependence in attentiveness to organized science (see Table 10.6). Since the level of career
and family interest are positively related, the relationship is additive.
    Both family and career interest are general in
nature, and speculative about the future. It is
perhaps this somewhat abstract nature of the variables that results in the absence of a relationship
with the present patterns of issue interest and attentiveness. While the prospective occupation or

profession of the respondents is also speculative
in nature, it is a more concrete decision than the
generalized expressions of life goals just reviewed.

Occupational Preference

The prospective occupation of the respondent was determined by asking each respondent to indicate his
or her expected occupation. Using the same set of
occupational classifications, the student reported
parental occupations (see Appendix A). Following
the analyses discussed in the introduction to this
section, the occupational preference information was
collapsed into a smaller number of categories and ultimately dichotomized into one set of occupations
that include scientific and engineering profession-
als and most public service careers, and all other
occupational preferences were clustered into a residual grouping.
     Using the multivariate log-linear model just
employed to assess family and career interests, and
leaving those measures in the model, the NPAS data
show that a preference for a scientific or public
service occupation is strongly and positively associated with attentiveness to organized science (see
Table 10.6). The direct effect of occupational preference in the logit model accounts for approximately
7 percent of the total mutual dependence in the model, which is a significant contribution to understanding the roots of attentiveness. An analysis of
higher-order interactions in the model disclosed no
higher-order terms that were significant at the
0.01 level or that explained more than 1 percent of
the total mutual dependence in the model.
     Looking at the analysis of the effects of the
three life goals and occupational preference measures, it appears that the major stimulus to attentiveness to organized science is the selection of a
specific career field related to science and technology. The contribution of an occupational preference
for a scientific or public service field in the preceding model is additive to the general level of
family and career commitment held by the respondent.
This result would suggest that one of the primary
motives for the development of attentiveness may
be the personal and professional interest of an individual in his or her career field, and not a more
general commitment to serving the public good

through the exercise of citizenship responsibilities
in this regard. This explanation will be explored
in great depth in later chapters focusing on the relationship of attentiveness to general political dispositions.

SUMMARY

It is now possible to specify school and peer influences in the development of attentiveness to
organized science within the context of basic home
and family influences. Several important patterns
have emerged that deserve emphasis.

First, the effect of educational plans and aspirations continues to be the most powerful predictor
of attentiveness to organized science in all of the
models examined in this chapter. The analysis in
the preceding chapter treated the development of college aspirations as primarily the result of family
socialization and the analyses of school and peer
variables do not show any enhancement or suppres-
sion of this relationship. While school and peer
experiences may motivate some individuals toward or
away from college, the NPAS data suggest that the
modal pattern is for high school and peer socialization to confirm and sustain the educational aspirations fostered by the student's home environment and,
perhaps, the elementary school years. These data
tend to dispute the image of the schools -- especially high schools -- as shapers of educational aspirations and sources of educational mobility.

Second, in a parallel manner, the preceding
analyses indicate that a strong gender difference
-- presumably rooted in early sex-role socialization -- persists without significant modification by
any of the school and peer experiences examined.
Despite the considerable public investment in career
education programs and the efforts of numerous
groups to encourage more women to seek scientific
and engineering careers, the NPAS data provide no
evidence that these efforts have had any significant
impact at the high school or college levels.

Third, the most effective sources of politicization toward science and technology issues are the
family and the peer group. The NPAS data indicate
that the frequency of classroom discussion of political issues bears no relationship to attentiveness
to organized science, but that both family and peer

discussion of political issues are positively and significantly related to attentiveness to organized science.

Fourth, attentiveness to organized science is strongly and positively related to expectations of a defined scientific, technological, or public service occupation or profession. Generalized interest in job-related achievement in a family was not significantly related to attentiveness. The NPAS data suggest that the origins of attentiveness may be rooted primarily in personal and job-related interests, rather than a more general commitment to citizenship duties. To the extent that similar motivations lead to the development of attentiveness to other policy areas, it would appear that attentive publics may be more reflective of Madison's "factions" than a focusing of basic citizenship interests.

11 Personality

The last two chapters have identified types of family and school experiences that are associated with attentiveness to science and technology. The present chapter will consider the types of people who are most likely to become attentive to science. Certain kinds of personalities are more compatible with the development of attentiveness than are others. This may be the case because students with certain kinds of personalities are more likely to take part in family and school activities that give rise to attentiveness. Or, it may be the case because certain kinds of family and school activities foster personalities that directly influence a person's propensity to become interested in science and a person's ability to comprehend scientific information.

PERSONALITY TRAITS ASSOCIATED WITH
ATTENTIVENESS TO SCIENCE

Attentiveness to science involves three components -- interest in science, knowledge about science, and regular exposure to science information. A person's personality may make him or her more or less disposed toward each of these components. Any personality traits that make a person more likely to become interested in science will, when combined with other factors that foster knowledge and exposure to science information, make attentiveness more likely.

Likewise, any personality traits that make it easier
for a person to learn about science or that dispose
a person toward becoming a regular consumer of science information will, when combined with interest,
make attentiveness more likely. What kinds of personality traits motivate a person to become interested in organized science or more open to science
information?

Self-esteem

One trait that might be important is self-confidence
or self-esteem. A number of facets of self-esteem
appear to foster interest in public issues in general and, for some, attentiveness to science and technology (cf. Sniderman 1975). For instance, perceiv-
ing oneself as a person who is capable of making up
one's own mind about public issues, who is listened
to by others, and who has the skills necessary to
influence what others think, might motivate a person
to keep informed about the kinds of public issues
people are talking about. On the other hand, per-
ceiving oneself as less competent to express oneself
to others and less popular or influential might make
it less likely that one would be particularly interested in keeping up with and discussing current
events.
     To determine whether these aspects of self-
esteem influence the extent to which students become
attentive to science and technology, the NPAS questionnaire included five items from Coopersmith's
(1967) measure of adolescent self-esteem:

1. When I express an opinion, others
    usually follow my suggestions.

2. I find it hard to talk in front
    of a group.

3. I usually give in very easily when
    people disagree with me.

4. I am popular with people my own age.

5. There are lots of things about me I
    would change if I could.

Respondents were asked to indicate the extent to
which each statement was "very much like me," "somewhat like me," or "not at all like me."
    Each item measures a somewhat different aspect
of self-esteem. A factor analysis of these five
items for the college subsample, based on the ordinal correlation coefficient gamma, evidences two underlying factors. This two factor structure replicates a pattern reported by Robinson and Shaver
(1973) indicating that self-derogation and leadership-popularity are two separate components of self-
esteem among students. In the present sample, these
two factors, rotated to oblimin criteria, are correlated -.42, suggesting a single, unified concept.
    Responses to the five self-esteem items are sig-
nificantly correlated with attentiveness to science
(see Table 11.1). The first three items in Table
11.1 reflect the self-derogation aspect of self-
esteem. The data indicate that students who report
difficulty speaking in front of a group, who give in
easily in the face of disagreement, and who would
like to change themselves are less likely to be attentive to organized science than are their more confident colleagues.
    The last two "self-esteem" items were designed
to tap the leadership-popularity dimension (Coleman
1961; Rosenberg 1965). The NPAS data indicate that
young people who perceive themselves as popular and
as opinion leaders are significantly more likely to
be attentive to science and technology issues than
are other students. This pattern suggests that
leadership and issue attentiveness are positively related, contradicting the image that students who are
attentive to public issues tend to be unpopular
"eggheads."
    To develop a single measure of each student's
self-esteem, the five items were combined into an
index that split the students into three distinct
groups:

  1. Those who clearly evidence positive
      feelings about themselves based on
      multiple sources of self-confidence
      (indicated by positive statements on
      three or more of the items and no more
      than one negative statement).

  2. Those who feel reasonably positive
      about themselves but harbor some self-
      doubts that reduce their general level

Table 11.1. Percent Attentive to Science by
Items Measuring Aspects of Self-esteem

Items	Not at all like me	Somewhat like me	Very much like me	Gamma
Hard to talk to groups	23 (1314)	15 (1537)	13 (980)	-.24
Gives in easily to others	21 (2354)	11 (1097)	9 (380)	-.36
Would change self	19 (964)	20 (1804)	12 (964)	-.15
Popular with own-age peers	10 (449)	17 (2208)	20 (1173)	.16
Others follow my suggestions	11 (390)	17 (2698)	22 (748)	.20

	Low	Moderate	High	Gamma
Self-esteem Index	8 (676)	16 (1843)	24 (1311)	.32

of self-confidence (as indicated by a mix of positive and negative responses to the five items).

3. Those who clearly have negative feelings about themselves in a number of domains (as indicated by negative statements on three or more of the items and no more than one positive statement.)

Using this index, 34 percent of the NPAS students evidenced relatively high self-esteem across the two domains measured by these items, and 24 percent of the high self-esteem group were attentive to organized science. On the other hand, 18 percent of the NPAS students scored low on the self-esteem index, and only 8 percent of this low-esteem group were attentive to organized science.

Political Efficacy

A second personality trait that might predispose a
person to become attentive to science and technology
issues is a person's feelings of political efficacy.
Gamson (1968), Zurcher and Monts (1972), and Paige
(1971) have argued that a general sense of political
efficacy is important to overt political participation. Watts (1973), however, has found conflicting
evidence. In view of the commitment of a person's
time and resources implied in the attentiveness construct, it is reasonable to assume that people who
see themselves and others as ineffective in the public domain are not likely to be very attentive to low
salience issues such as science and technology.
    In the work discussed above and in the classic
political socialization studies of Hess and Torney
(1967) and Easton and Dennis (1969), the concept of
political efficacy was broad in nature, referring to
individuals' views of the political impact of participation by the modal citizen. While an understanding of this more general attitude is important, it is
possible that people make a finer distinction when
considering what impact citizens who are well informed on an issue might have. It is also possible, of
course, that those who view the political efficacy
of the informed citizen as greater might be more motivated to become attentive to public issues themselves.
    Recognizing the imprecision of the traditional
definition, the NPAS questionnaire included a set of
four items that asked the respondent to express his
or her agreement or disagreement with the statement,
"The interested and informed citizen can often have
some influence on ... policy decisions if he is willing to make the effort." The data demonstrate that
the young adults in the NPAS sample made important
differentiations among the four policy areas in the
level of political influence available to interested
and informed citizens. Eighty-five percent of the
respondents who either agreed or disagreed with the
statements indicated that attentive citizens could
influence civil rights policies, compared to 67 percent who felt that an interested and informed citizen
could significantly impact foreign policy decisions
(see Table 11.2). In between these two figures, 73
percent of the respondents expressing an attitude
agreed that the interested and informed citizen
could have an influence on economic policy, and 74

Table 11.2. Percent Attentive to Science by
Perceived Efficacy in Selected Policy Areas

"The interested and informed citizen can often have some influence on...	SD	D	A	SA	Gamma
Science & technology policy	13% (137)	17% (734)	18% (2164)	20% (350)	.07
Foreign policy	19 (265)	19 (819)	18 (1791)	15 (449)	-.06
Economic policy	9 (136)	21 (759)	17 (2026)	19 (355)	-.01
Civil rights policy	10 (101)	14 (414)	19 (2278)	20 (632)	.12

percent indicated they felt that science and technology policies could be influenced by attentive citizens.

While the level of perceived attentive efficacy
differs significantly within the NPAS population,
attentiveness to organized science appears to be
only weakly related to perceived efficacy in any of
the four areas. The strongest association between
attentiveness and political efficacy is found in
regard to the question concerning civil rights, but
the gamma coefficient of 0.12 signals a very weak
relationship. The virtual absence of any relationship between perceptions of efficacy in regard to
foreign or economic policy and attentiveness to
science is supportive of the general issue specialization thesis outlined earlier.

In view of the previous literature concerning
general political efficacy and the findings of a
relationship to political participation, the four
judgments about the political efficacy of attentive
citizens were combined to construct an index of political efficacy for interested and informed citizens.
Respondents who agreed that attentive citizens could
be influential in three or more of the issue domains

were classified as high on the index of political
efficacy and the residual was coded as low. The
dichotomous index proved to be a weak predictor of
attentiveness to organized science, producing an ordinal correlation of only 0.15, which is not a significant improvement over the predictive power of
the individual items.

Open- and Closed-mindedness

The NPAS data reviewed above suggest that feelings
of self-esteem influence young adults to become
interested in public issues in general and science
and technology policies in particular. But attentiveness involves more than interest. It also
involves knowledge. What kinds of people are more
likely to become knowledgeable about science and
technology during their high school and college
years?
     The last chapter reported that students' academ-
ic competence, as measured by the students' reported
rank in their respective schools, made some difference in the level of cognitive knowledge different
students hold about science and technology.
     Relative academic competence, however, is not
the only personal characteristic that might contribute to the comprehension and retention of information about science and scientific issues. Besides
being able to comprehend technical information about
about science, the student must also be willing to
consider different points of view.
     Public issues are frequently controversial.
There are rarely clear-cut, unambiguous answers or
simple solutions to problems. To comprehend the
science and technology information available in the
mass media, a person must be able to deal with new
concepts and new ways of thinking about problems.
One must be able to approach issues in a somewhat
open-minded manner that allows the consideration of
information that is not entirely consistent with
previously held information on which one has based
existing attitudes. The person who is incapable of
adopting new ways of thinking or is unwilling to try
to incorporate new ideas and new approaches is not
likely to comprehend much information about science
and technology issues even when he or she is exposed
to it.

To test the validity of these arguments, the
NPAS questionnaire included a series of indices
designed to measure open- and closed-mindedness to
public issues. These indices reflect work done by
Milton Rokeach in the 1950s when he developed and
tested measures of dogmatic thinking, published in
his book entitled The Open and Closed Mind (1960).
According to Rokeach's studies, open-mindedness
involves a number of distinct characteristics. First,
to be open-minded one must be willing to listen to
and consider other people's ideas. One cannot be
estranged from others nor feel alienated from what
others believe. Any beliefs or attitudes that push
other people away and separate them and their ideas
as alien, foster the development of a closed-minded,
dogmatic way of thinking.
Second, to be open-minded one must be able to
entertain a variety of approaches to problems without a tendency to settle on a single solution pre-
maturely and close out potentially useful alternatives. Open-mindedness, therefore, requires a toleration of potentially conflicting ideas and avoidance
of the kind of single-minded approach to problems
that cannot be altered by new experiences (cf. Hoffer's 1958, The True Believer).
Third, open-mindedness involves the use of
objective criteria to judge the value and validity
of new ideas. The tendency to judge ideas on the
basis of irrelevant characteristics of the person or
agency providing the information signals the existence of a closed-minded approach to ideas. The
open-minded person, therefore, is less likely than
the closed-minded to be ethnocentric in an evaluation
of the worth of ideas and is more willing to entertain the possibility that people with different philosophies are sometimes correct in what they believe.
Rokeach developed a 66-item scale to measure
the multiple dimensions of open- and closed-
mindedness. The NPAS study selected ten items from
this scale that, on the basis of previous studies
(Suchner 1977) , fit a pattern suggesting the validity of Rokeach's characterization of the open- and
closed-minded person. The following Rokeach items,
to which each respondent was asked to respond from
"strongly agree" through "strongly disagree," were
included in the 1978 NPAS:

1. Fundamentally, the world we live in is
    a pretty lonesome place.

2. Most people just don't give a damn for others.

3. Of all the philosophies in the world, there is probably only one which is correct.

4. When it comes to differences of opinion, we must be careful not to compromise with those who believe differently than we do.

5. It is only natural that a person should have a much better acquaintance with ideas he believes in than ideas he opposes.

6. In this complicated world of ours, the only way we can know what is going on is to rely on leaders and experts who can be trusted.

7. In the long run, the best way to live is to pick friends and associates whose tastes and beliefs are the same as your own.

8. A person who has not believed in a great cause has not really lived.

9. It is only when a person devotes himself to an ideal or cause that life becomes meaningful.

Respondents were also asked to answer "not at all like me," "somewhat like me," or "very much like me" in response to the statement:

10. If given a chance, I would do something of great benefit to the world.

A factor analysis of these ten items for the college subsample, based on a gamma matrix, produced a three factor structure consistent with Rokeach's previous findings (see Table 11.3).(1) The factors were rotated to an oblique solution to allow the most direct measurement of the correlated facets of the open and closed mind. There is a 0.25 correlation between Factor I and Factor II, a -.28 correlation between Factor I and Factor III, and a -.16 correlation between Factor II and Factor III, indicating that the Factors II and III share more

Table 11.3. Factor Pattern for Items Measuring
Open- and Closed-mindedness in College Subsample

Items	I	II	III	h²
Fundamentally, the world we live in is a pretty lonesome place.	.48	-.10	-.04	.23
Most people just don't give a damn for others.	.63	-.05	.08	.36
Of all the philosophies in the world, there is probably only one which is correct.	.56	.14	-.17	.47
When it comes to differences in opinion, we must be careful not to compromise with those who believe differently than we do.	.34	.23	-.10	.25
It is only natural that a person should have a much better acquaintance with ideas he believes in than ideas he opposes.	-.04	.53	.05	.27
In this complicated world of ours, the only way we can know what is going on is to rely on leaders and experts who can be trusted.	-.11	.53	-.07	.27
In the long run, the best way to live is to pick friends and associates whose tastes and beliefs are the same as your own.	.20	.40	.04	.24
A person who has not believed in a great cause has not really lived.	.12	.11	-.75	.67

Table 11.3 (continued). Factor Pattern for Items
     Measuring Open- and Closed-mindedness
              in College Subsample

Items	Factors			2
	I	II	III	h
It is only when a person devotes himself or herself to an ideal or cause that life becomes meaningful.	.16	.16	-.59	.50
If given a chance, I would do something of great benefit to the world.	-.05	-.08	-.35	.12

| Proportion of common variation accounted for: | | | | Total |
| | .27 | .14 | .13 | .54 |

Note: Factor pattern based on analysis of gamma
        correlation coefficients and an oblique
        rotation of the largest factors to the
        oblimin solution.

common content with Factor I than they do with each
other.
        Factor I is best defined by the first four
items. Taken together, these four items appear to
tap a sense of estrangement and aloofness from other
people and their ideas. Respondents who agreed that
"most people just don't give a damn for others" were
also quite likely to indicate a sense of estrangement from others in their feelings of loneliness and
in their caution against considering others'
opinions and philosophies. These four items can be
used, therefore, to create an index of the extent
to which the respondents feel "estranged" from other
people and their ideas.
        The next three items all appear to reflect the
idea that one is best off if other people are
separated into those with shared beliefs and those
with conflicting beliefs. The function of such
division of the world into "we" and "they" appears
to be to simplify what otherwise might seem to be a
complicated, unfamiliar, and possibly frightening

world of ideas. The correlation of 0.25 between the
first and second factors suggests some commonality
between the "estrangement" indicated in the first
four items and the segregation of peoples and their
ideas into distinct camps in the second set of three
items. The content that distinguishes these two
factors appears to focus on the implications of
living in an alienated world, a primary implication
being that one should be cautious in choosing "similar" friends and in listening only to "trusted"
leaders. Following Rokeach's (1960) characterization of the closed-minded person as relatively
ethnocentric in the criteria used to separate the
world into "we" and "they," Factor II can be interpreted as indexing the degree to which respondents
tend to focus their attention on trusted members of
their own group and tend to distrust and ignore information provided by people dissimilar from themselves.

The last three items each appear to be related
to endorsement of a single-minded concern for discovering and pursuing a cause that makes life
"meaningful." This "true-believer" approach to life
is clearly distinguished from the other two dimensions of closed-mindedness identified as common
among the other two factors, but the correlations
of Factors I and II with Factor III indicate that
people who tend to be estranged from others and
their ideas and people who tend to be oriented
toward members of their own group tend not to be
those who take a rather single-minded approach to
ideas that make "causes" particularly attractive.

There is, however, a distinction that might
be made between the first two items and the last
item loading on Factor III, which appears to be
uncorrelated with Factors I and II. The first two
items are prescriptive -- one ought to believe and
devote oneself to a cause. The last item, on the
other hand, is predictive -- the student reports he
or she will do something of great benefit, if given
the chance. The smaller loading of the third item
on Factor III suggests that the underlying commonality among these three items has less to do with
prediction than with prescription of a cause-
oriented, single-minded approach to life.

Given the multidimensional nature of open- and
closed-mindedness, three indices were constructed to
measure estrangement, ethnocentrism, and single-
mindedness among the students in the NPAS sample
using the items best measuring each factor. Each

index split the sample into two groups: (1) those
who evidenced closed-minded tendencies through
agreement with a majority of items measuring each
factor and (2) those who evidenced open-minded
tendencies by disagreement or who evidenced little
tendency in either direction.

Estrangement

Each of the four items measuring estrangement from
people and their ideas is negatively correlated with
attentiveness to science, as is the combined index
(see Table 11.4). Students who perceive others as
not caring or who express caution about considering
others' opinions and philosophies are not likely to
be attentive to science. Of the students who agreed
with a majority of the "estrangement" statements,
only 11 percent are attentive to science. On the
other hand, 21 percent of the students who disagreed
or expressed uncertainty regarding a majority of
these statements qualify as attentive to organized
science.

Ethnocentrism

Two of the three items measuring ethnocentric tendencies evidence small negative correlations with
attentiveness, as does the index of ethnocentrism
(see Table 11.4). Students who would rely primarily
on "trusted leaders" for their information and those
who would try to choose friends with similar opinions and tastes as their own apparently lack some of
the curiosity, toleration, and possibly the self-
confidence that fosters attentiveness to public
issues. The association between ethnocentric tendencies and attentiveness to organized science is negative and weak, suggesting that, for this age-group
at least, orientations toward other people has less
to do with the development of interest and knowledge
than orientations toward ideas and philosophies.

Single-mindedness

Two of the three items measuring "single-minded"
tendencies to look for and endorse causes as a
meaningful way to live show negative correlations
with attentiveness, while the third item shows a
positive correlation (see Table 11.4). As suggested above, the distinction among these items appears
to be between prescribing that one ought to believe

Table 11.4. Percent Attentive to Science by
Items and Indexes Measuring Open- and
Closed-Mindedness

Factor I: Estrangement					

Items	SD	D	U	A	SA	Gamma
The world is a lonesome place	19% (507)	19% (1964)	12% (354)	16% (916)	12% (151)	-.10
People don't give a damn	21 (318)	22 (1498)	14 (275)	15 (1257)	9 (575)	-.24
Only one philosophy is correct	25 (747)	19 (1638)	10 (707)	8 (565)	16 (172)	-.30
Risky to compromise ideas	31 (452)	22 (1660)	9 (498)	12 (1037)	18 (224)	-.22

	Low	High	Gamma
Estrangement Index	21 (2404)	11 (1490)	-.34

===

Factor II: Ethnocentrism					

Items	SD	D	U	A	SA	Gamma
Natural to know own ideas better	19 (134)	21 (790)	9 (460)	16 (2082)	21 (405)	.00
Must rely on leaders to know what's going on	20 (332)	20 (1431)	9 (516)	16 (1373)	15 (219)	-.10
Best to pick friends similar to self	14 (266)	20 (1433)	17 (389)	15 (1470)	18 (333)	-.07

	Low	High	Gamma
Ethnocentrism Index	19 (1897)	16 (1974)	-.10

Table 11.4 (continued). Percent Attentive to
Science by Items and Indexes Measuring
Open- and Closed-Mindedness

Factor III: Single-mindedness

Items	SD	D	U	A	SA	Gamma
Must believe in cause	22 (310)	21 (1693)	13 (702)	13 (971)	11 (194)	-.21
Must devote to cause	18 (131)	23 (947)	14 (404)	15 (1825)	17 (565)	-.11

	Not at all like me	Somewhat like me	Very much like me	Gamma
Would do something to benefit world if given chance	11 (695)	16 (1620)	22 (1485)	.23

	Low	High	Gamma
Single-mindedness	18 (2283)	16 (1588)	-.04

in a great cause and predicting that one will do
something of great benefit for the world. While
respondents who agreed with the prescriptive statements were also likely to agree with the predictive
statement, these two tendencies appear to have
different implications for becoming attentive to
science. Believing one should devote oneself to a
cause to "make life meaningful" or to "really live"
is negatively related to attentiveness to science
and technology issues. This suggests that organized
science is not likely to be associated with identi-
fiable "causes" in the minds of most high school and
college students. On the other hand, predicting
that one will do something of great benefit to the
world, if given the chance, is positively associated
with attentiveness to science and technology. This
suggests that science and technology represent, to
high school and college students, possible avenues
to make contributions to society.

The combination of positive and negative associations with attentiveness among items measuring the same general construct -- single-mindedness -- attenuates the association between the composite index and attentiveness. The correlation of -.04 between the index of single-mindedness and attentiveness is smaller than any of the three correlations between the individual items and attentiveness. This occurs because some students both believe in the worth of causes and predict that they will do some great benefit for the world, and these beliefs apparently tend to balance off each other in their effects on developing attentiveness to science.

Combining Estrangement, Ethnocentrism,
and Single-mindedness

The correlations among the three factors of closed-mindedness suggest that these dimensions may not have independent associations with attentiveness. To assess this possibility, the three dimensions were cross-classified and used in predicting attentiveness (see Table 11.5).

The logit model for these data indicates that of the three dimensions of closed-mindedness, only estrangement has a residual bivariate association with attentiveness to organized science (see Table 11.6). Apparently, ethnocentrism is associated with attentiveness only through an interaction with estrangement, and single-mindedness appears to have no significant residual association with attentiveness once these other two dimensions of closed-mindedness have been taken into account. Young adults who measure low on the estrangement index are consistently more likely to be attentive to science than are those who measure high on estrangement, regardless of the level of ethnocentrism. Among young adults measuring high on estrangement, however, ethnocentrism further reduces the probability of attentiveness (see Fig. 11.1).

It is apparent, then, that two dimensions of closed-mindedness play some role in determining whether a student's personality is compatible with the development of attentiveness to organized science. If a student is both estranged from other people and their ideas and ethnocentric in his or her criteria for judging the worth of others' ideas, it is not likely the student will become attentive

Table 11.5. Percent Attentive to Science by
Level of Estrangement, Ethnocentrism,
and Single-Mindedness

Single-mindedness	Ethno-centrism	Estrange-ment	Attentive to Science	
			N	%
Low	Low	Low	968	21
		High	302	12
	High	Low	651	19
		High	362	9
High	Low	Low	366	16
		High	257	17
	High	Low	404	24
		High	557	10
Total			3867	17

to science and technology issues. If, on the other
hand, the student is open to other people and their
ideas and is willing to consider ideas and philosophies that are unfamiliar, the likelihood of becoming attentive to science increases.

Faith in People

Another personality characteristic one might suspect correlates with attentiveness to science is the
degree to which students trust and have faith in people. Numerous studies demonstrate that for people
to be influenced by information from others, they
need to perceive the source of the information as
both competent and trustworthy (cf. Kelman 1958;
Walster et al. 1966). If an individual believes that
other people are generally untrustworthy or devious
in what they report about the discoveries and ad-
vances of science and technology, he or she is not
likely to find the information very valuable and is
likely to pay little attention to what others say
and write about scientific issues. To be attentive
to organized science, therefore, one would suspect
that besides being open to others in general (as
suggested by the correlation between estrangement
and attentiveness), a person needs to place at least
some trust in other people.

236   CITIZENSHIP IN AN AGE OF SCIENCE

Table 11.6.   Models for Data in Table 11.5

Models	LRX^2	df	CMPD
H1 EIS,A	75.99	7	---
H2 EIS,AE,AI,AS	17.74	4	.77
H3 Model H2 minus AE	70.30	5	---
Difference due to AE	52.57	1	.69
H4 Model H2 minus AI	18.87	5	---
Difference due to AI	1.14	1	.02
H5 Model H2 minus AS	18.21	5	---
Difference due to AS	.47	1	.01
H6 EIS,AE,AI,AS,AEI	10.27	3	.86
H7 Model H6 minus AEI	17.74	4	---
Difference due to AEI	7.46	1	.10

A = attentive to science
E = estrangement
I = ethnocentrism
S = single-mindedness

To test this hypothesis, three items first used by Rosenberg (1957) were adopted to measure the respondents' trust in people and to investigate the relationship between social trust and attentiveness to organized science. The three items are:

1. Would you say that most of the time people try to be helpful, or that they are mostly looking out for themselves?

2. Do you think that most people try to take advantage of you if they get the chance, or do most people try to be fair?

3. Generally speaking, would you say that most people can be trusted, or that you can't be too careful in dealing with people?

On all three items, those young adults who are
less trusting and have less faith in others are also
less likely to be attentive to organized science
(see Table 11.7). Twenty percent of the students
who indicated they think people are generally trustworthy, helpful, or fair are attentive to science,
compared to 14 percent of the students who indicated
they think people are generally untrustworthy or out
for themselves. Those students who indicated they
had "no opinion" on each of the individual items
were the least likely to be attentive to science.
These data suggest that having not made up one's
mind about whether one can trust what others say is
just as detrimental to the development of attentiveness as having developed a cautious, or even a
cynical, viewpoint on the credibility of others'

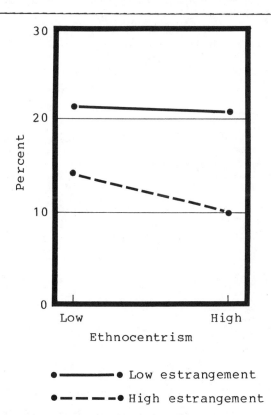

Fig. 11.1. Percentage attentive to organized science
by level of estrangement and level of ethnocentrism.

Table 11.7. Percent Attentive to Science by
Items Measuring Faith and Trust in People

"People tend to . . ."

Be out for themselves	No opinion	Be helpful	Gamma
17%	8%	21%	.09
(2106)	(594)	(1305)	

Take advantage	No opinion	Be fair	Gamma
15	11	20	.18
(1747)	(475)	(1784)	

"Can most people be trusted?"

You can't be too careful	No opinion	Most people can be trusted	Gamma
14	12	24	.26
(2379)	(451)	(1176)	

	Low	High	Gamma
Trust in People Index	14	20	.20
	(2114)	(1892)	

information. Attentiveness to organized science
apparently involves making a commitment to listen to
others and to make up one's own mind. It does not
necessarily mean that one must trust everything one
hears or reads.

Relationships Among the Personality Traits

The analysis in the preceding sections points toward a number of personality traits that are related
to attentiveness to science and technology issues.
These personality traits are, however, often correla-

ted with each other. Students who are trusting of
others are also more open to other's ideas and philosophies. Similarly, students who are more open to
others' ideas tend to evidence higher levels of
self-esteem. To determine which personality traits
are central to the development of attentiveness,
then, it is necessary to examine the residual association of each trait with attentiveness to science
within subsamples of respondents who are similar on
the other dimensions in the analysis (see Table
11.8).

The residual associations among the personality
traits indicate that estrangement and trust are, by
far, the most strongly related personality traits
(see Fig. 11.2). Students who do not trust other
people are very likely to express estrangement from
other people's ideas and philosophies. Estrangement
is also residually associated with self-esteem. Students who have higher self-esteem are somewhat less
likely to be estranged from other people.

Perceived efficacy of informed citizens is less
strongly associated with the cluster of self-esteem,
trust, and estrangement. Students who express estrangement from others are less likely to believe
individual action is effective in changing public
policy. However, holding constant self-esteem and
trust in people, the relationship between feelings
of estrangement and cynicism regarding the effectiveness of individual action does not seem as strong as
one might have expected.

Self-esteem makes the strongest contribution in
predicting attentiveness to science independent of
the influence of the four other traits. Students
who feel confident about their own opinions and who
are comfortable about expressing these opinions are
more likely than other students to develop an interest and attentiveness to science. The present
analysis suggests, however, that when these feelings
of self-esteem are combined with an open mind toward
other people and their ideas, with a feeling of efficacy in individual action, and with a trust in
people, the likelihood of attentiveness is increased. Of the young adults in the NPAS sample who
lacked all four of these characteristics, only 11
percent were attentive to science, while 26 percent
of the young adults who measured high on all four
characteristics were attentive to science.

Table 11.8. Percent Attentive to Science by
Self-esteem, Trust in People, Estrangement,
Political Efficacy

Political Efficacy	Estrange-ment	Trust	Self-esteem	Attentive to Science N	%
Lower	Lower	Lower	Low	71	11
			Medium	219	14
			High	176	22
		Higher	Low	75	9
			Medium	267	17
			High	176	22
	Higher	Lower	Low	86	12
			Medium	182	11
			High	114	14
		Higher	Low	53	6
			Medium	75	4
			High	37	14
Higher	Lower	Lower	Low	75	7
			Medium	272	15
			High	201	29
		Higher	Low	86	13
			Medium	350	30
			High	269	28
	Higher	Lower	Low	132	5
			Medium	256	11
			High	164	14
		Higher	Low	60	10
			Medium	130	13
			High	78	26
Total				3661	17

INTEREST, KNOWLEDGE, AND EXPOSURE TO
SCIENCE INFORMATION

The preceding analysis has indicated the kinds of
personality traits that differentiate students who
are attentive to organized science from those who
are not. As pointed out in the discussion of the
association of academic competence with attentiveness, however, these correlations do not indicate
whether a personality trait is associated with
attentiveness primarily through its association with
interest in science, with knowledge about science,
or with regularity of exposure to science information. Hence, it is difficult to infer why students
with different personality dispositions evidence
different levels of attentiveness or to develop
notions about the kinds of changes in personality
that would alter the level of a student's attentiveness to public issues.
     To determine the relationship of each personality
trait with each component of attentiveness, it is necessary to examine the personalities of students who
combine the components of attentiveness in different
ways. By looking at the conditional association of
each personality trait with each component of attentiveness, one can identify which personality traits
are associated with interest in science, knowledge
about science, and exposure to science information.
     The simplest approach to comparing the person-
ality traits of students who combine different aspects of attentiveness to science and technology is
to partition the NPAS sample into eight subsamples.
One subsample is made up of students who combine all
three components of attentiveness (i.e., they are interested in science, knowledgeable about science,
and regular consumers of science information). In
the preceding analysis, these students have been
labeled "attentive to organized science." The other
seven subsamples are made up of students who are less
attentive to science due to the absence of one, two,
or all three of these characteristics.(2)
     The NPAS data indicate that 17 percent of the
students combine all three components of attentiveness -- interest, knowledge, and regular information
consumption -- that constitute attentiveness to organized science or technology (see Table 11.9). The
remaining 73 percent evidence lack of attentiveness
in one or more of the components. Thirty-five percent are not interested in science or technology; 41

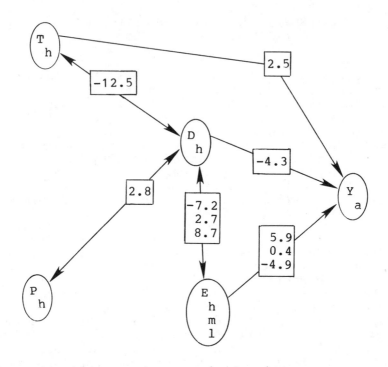

Y = attentiveness to organized science
 a   attentive

T = level of trust in people
 h   high

P = level of perceived political efficacy
 h   high

D = level of estrangement
 h   high

E = level of self-esteem
 h   high
 m   medium
 l   low

Fig. 11.2. A path model of the relationship between
attentiveness to organized science and selected
personality attributes.

percent are not knowledgeable about science or technology; and 75 percent do not participate regularly
in an information network providing information
about science and technology issues.

The strong positive relationships among the
three components of attentiveness mean that personality differences that are associated with one of
the components of attentiveness are likely to be
associated with the other two components as well.
Part of this association will be due to commonalities in the factors that influence both the development of the personality traits and the development
of attentiveness, while part of this association will
be due to the influence of one component of attentiveness on another.

The NPAS data indicate that self-esteem is
strongest among students scoring high on all three
attentiveness components (see row one of Table
11.9). The logit model that fits these data indicates signficant conditional relationships between
self-esteem and each of the three components of attentiveness (see Table 11.10). The strongest associations, however, are between self-esteem and the
knowledge and information consumption components.
Students who are knowledgeable about science and
technology and regular consumers of science information are about twice as likely to evidence high self-
esteem as are students who are neither knowledgeable
nor regular information consumers.

In the area of perceived political efficacy,
the NPAS data indicate that a higher proportion of
the attentives display this trait than nonattentives
(see row two of Table 11.9). The logit model that
fits these data indicates that perceived political
efficacy is, however, conditionally associated with
only one of the three components of attentiveness
(see Table 11.11). Students who are optimistic
about the effectiveness of individual action in the
public arena are more likely to be knowledgeable
about science and technology issues than are students who are less optimistic about the efficacy of
individual action. It would appear, then, that perceived political efficacy has little direct association with students' interests or information acquisition habits regarding science and technology.

In reference to the three indices of open-
mindedness, the NPAS data suggest that a sense of
estrangement and ethnocentrism are both negatively
associated with attentiveness, but that the cause
orientation index is largely unrelated to the atten-

Table 11.9. Percent Measuring High on each
Personality Index among Students Evidencing
Different Combinations of the Components of
Attentiveness to Science and Technology

Trait	None	I	K	C	IK	IC	KC	IK	Total
Self-esteem	25%	29%	30%	29%	37%	34%	45%	47%	34%
Political efficacy	49	52	54	58	60	47	54	62	55
Estrange-ment	49	56	31	49	30	55	30	25	38
Ethno-centrism	54	54	49	50	50	61	46	47	51
Cause ori-entation	41	47	40	33	40	51	31	39	41
Trust in people	42	42	45	43	49	46	61	56	47
N	781	643	448	60	1123	148	133	670	4007

I = interest in organized science
K = knowledge of science
C = consumption of science/technology information

Note: The number of respondents in each subsample
is based on the total N of 4007 as the best
estimate of the proportions of students in
each subpopulation. The actual N's on which
the percentages in the table are based range
from 3807 to 4007.

tiveness cluster (see rows three, four, and five of
Table 11.9).   Twenty-five percent of the students
who are attentive to science on all three components
of attentiveness evidence low levels of estrangement
compared to 49 percent of students who are not attentive on any of the three components. The differences
in representation between these two extreme groups on
the other two dimensions of closed-mindedness are

considerably smaller. It is apparent that estrangement from people and their ideas can be a major hin-
drance in developing familiarity with the basic concepts of science and with science issues. The logit
models that fit the data for the estrangement and
ethnocentrism dimensions of open-mindedness are
strongly associated only with the knowledge component
of attentiveness (see Table 11.12).

Finally, the NPAS data indicate that trust in
people is positively associated with attentiveness
to organized science (see last row of Table 11.9).
The logit model that fits these data shows that
trust in people is associated with both knowledge
about science and regularity of information acquisition (see Table 11.13). Students who are knowledgeable about science and technology and students who
expose themselves to information about science and
technology on a regular basis are more likely to indicate they trust what other people tell them. An
estimated 57 percent of the students who combine
both knowledge and regular consumption of scientific
information evidence a high level of trust in people,

Table 11.10. Models of Self-esteem Predicting
the Components of Attentiveness
to Science and Technology

Models		LRX^2	df	CMPD
H1	IKC,S	170.73	14	---
H2	IKC,SI,SK,SC	12.86	8	.92
H3	Model H2 minus SI	29.73	10	---
	Difference due to SI	16.87	2	.10
H4	Model H2 minus SK	63.55	10	---
	Difference due to SK	50.69	2	.30
H5	Model H2 minus SC	46.21	10	---
	Difference due to SC	33.35	2	.20

I = interest in organized science
K = knowledge of science
C = consumption of science/technology information
S = self-esteem

Table 11.11. Models of Perceived Efficacy
Predicting the Components of Attentiveness
to Science and Technology

Models	LRX^2	df	CMPD
H1 IKC,E	38.79	7	---
H2 IKC,EI,EK,EC	5.26	1	.86
H3 Model H2 minus EI	9.90	5	---
Difference due to EI	4.64	1	.12
H4 Model H2 minus EK	23.58	5	---
Difference due to EK	18.32	1	.47
H5 Model H2 minus EC	5.58	5	---
Difference due to EC	.31	1	.01

I = interest in organized science
K = knowledge of science
C = consumption of science/technology information
E = perceived efficacy

compared to an estimated 42 percent of the students
who indicate neither knowledge nor regular exposure.

A MODEL OF PERSONALITY FACTORS ASSOCIATED
WITH ATTENTIVENESS TO SCIENCE

The preceding analyses have attempted to accom-
plish three tasks. First, the associations of
each of the items measuring different aspects of
attentiveness to science and technology have been
examined. Out of that analysis, indices measuring
a number of core concepts of personality predictive
of attentiveness were constructed and interpretations were given indicating which items contributed
the most to differentiating between students who are
attentive to science and those who are not.
     The second part of the analysis combined the in-
dividual personality traits to determine how associations among the personality traits confound variation
in their association with attentiveness to science.

This analysis determined which of the personality
traits best predict attentiveness and which personality traits are associated primarily through association with other traits.

The third part of the analysis partitioned the
concept of "attentiveness to science" into its three
components -- interest, knowledge, and information
consumption -- to determine which of the personality
traits is associated with which components. This
analysis distinguished between those personality
traits that have direct associations with each component and those traits that are associated with each
component primarily through association with other
components of attentiveness.

The results of these three analyses can be
summarized in a path diagram relating each of the
personality traits to each of the components of
attentiveness (see Fig. 11.3). There are strong
associations among each of the three components of
attentiveness to science. Assuming that some of
this association is due to direct causal relationships among the components, one can conclude that
some of the empirical association between the personality traits and each component is due to the
causal effects of the personality trait on a particular component followed by the transmission of
these personality effects throughout the structure
of attentiveness. Self-esteem, for instance, is
positively correlated with knowledge of science and
technology in part through its direct causal effects
on interest and information and the effects of each
of these components on the development of knowledge
about science.

Another part of the association between self-
esteem and attentiveness is apparently due to the
association of self-esteem with other personality
traits that directly predict components of attentiveness to science. In particular, the association
of self-esteem with estrangement accounts for most
of the remainder of the association between self-
esteem and the knowledge component of attentiveness.

Looking at the associations among the personality traits, it is apparent that political efficacy
of informed citizens is positively associated with
estrangement and that estrangement is, in turn, negatively associated with trust in people. These
personality traits, however, appear to affect attentiveness in different ways. Political efficacy
evidences a positive association with the development of knowledge about science, while estrangement

Table 11.12. Models of Dimensions of Closed-
mindedness Predicting the Components of
Attentiveness to Science and Technology

Factor I: Estrangement from other people
and their ideas

Models	LRX^2	df	CMPD
H1 IKC,E	235.35	7	---
H2 IKC,EI,EK,EC	8.64	4	.96
H3 Model H2 minus EI	10.07	5	---
Difference due to EI	1.42	1	.01
H4 Model H2 minus EK	204.12	5	---
Difference due to EK	195.48	1	.83
H5 Model H2 minus EC	11.69	5	---
Difference due to EC	3.05	1	.01

Factor II: Ethnocentrism--oriented toward
own trusted group

Models	LRX^2	df	CMPD
H1 IKC,G	17.17	7	---
H2 IKC,GI,GK,GC	3.69	4	.79
H3 Model H2 minus GI	4.24	5	---
Difference due to GI	.55	1	.03
H4 Model H2 minus GK	15.41	5	---
Difference due to GK	11.72	1	.68
H5 Model H2 minus GC	4.31	5	---
Difference due to GC	.62	1	.04

Table 11.12 (continued). Models of Dimensions of
Closed-mindedness Predicting the Components of
Attentiveness to Science and Technology

Factor III: Single-mindedness--oriented
toward causes

Models		LRX^2	df	CMPD
H1	IKC,L	21.84	7	---
H2	IKC,LI,LK,LC	7.08	4	.68
H3	Model H2 minus LI	14.21	5	---
	Difference due to LI	7.13	1	.33
H4	Model H2 minus LK	16.65	5	---
	Difference due to LK	9.58	1	.44
H5	Model H2 minus LC	7.82	5	---
	Difference due to LC	.74	1	.03

I = interest in organized science
K = knowledge of science
C = consumption of science/technology information
E = estrangement
G = ethnocentrism
L = single-mindedness

is negatively associated with the development of
knowledge about science. Students who perceive the
possibility that the informed citizen can affect public policy and students who are open to other people
are more likely to become informed about public issues, including those dealing with science and technology. Students who do not believe the individual
can be effective or who express caution in dealing
with people and ideas that are foreign to them are
less likely to become knowledgeable in areas others
can inform them about.

Trust in people appears to have a direct positive association with exposure to information about
science and technology in the mass media. Students
who express trust in what people say, and faith that
other people are likely to tell the truth most of

Table 11.13. Models of Trust in People Predicting
             the Components of Attentiveness
             to Science and Technology

Models	LRX^2	df	CMPD
H1 IKC,T	48.82	7	---
H2 IKC,TI,TK,TC	4.57	4	.91
H3 Model H2 minus TI	5.05	5	---
Difference due to TI	.47	1	.01
H4 Model H2 minus TK	20.11	5	---
Difference due to TK	15.54	1	.32
H5 Model H2 minus TC	19.03	5	---
Difference due to TC	14.46	1	.30

I = interest in organized science
K = knowledge of science
C = consumption of science/technology information
T = trust in people

the time are more likely to pay attention to what
others write in the mass media, including what they
write about science and technology.
     The personality syndrome that connects lack of
estrangement from others with faith in people and
perceived efficacy of individual actions appears to
play an important role in the development of attentiveness through effects on each of the three components of attentiveness. When combined with positive
self-esteem, it is apparent that the student who
exhibits all of these characteristics is quite likely to become attentive to science and technology,
while the student who exhibits none of these characteristics is very unlikely to become attentive.

                Combining Personality with School
                    and Family Effects

The personality theory illustrated in Figure 11.3
provides a basis for predicting what kinds of stu-

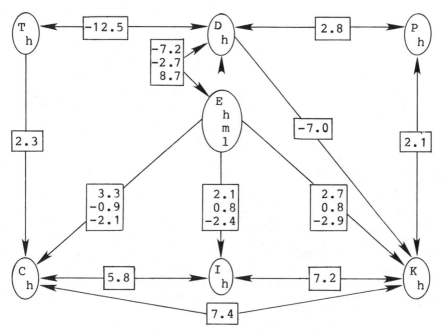

K = level of knowledge about organized science
 h   high

I = level of interest in organized science issues
 h   high

C = level of science information consumption
 h   high

T = level of trust in people
 h   high

D = level of estrangement
 h   high

P = level of perceived political efficacy
 h   high

E = level of self-esteem
 h   high
 m   medium
 l   low

Fig. 11.3. A path model of the relationships among
the components of the attentiveness typology and
selected personality attributes.

dents will become attentive to science and technology. Combining this personality theory with the
social factors that affect the development of
attentiveness raises questions about how these
social factors are associated with the personalities
of the students and how these associations make it
more or less likely that they will become attentive
to science.

Personality syndromes and social factors
combine to affect social processes in a number of
ways. First, social factors associated with different families and schools can influence the kinds
of personalities students develop. If a student
lives in a home environment that fosters the development of high self-esteem, trust in people, and
open-mindedness toward others' ideas, and if the
school the student goes to reinforces these tendencies, the student is likely to develop the kind of
personality that is associated with attentiveness to
science. On the other hand, if the student lives in
a home environment that undermines the development
of self-esteem and creates a distrusting, closed-
minded disposition toward other people and their
ideas, and if the school environment does little to
change these dispositions, the student is less
likely to develop the kind of personality that is
associated with attentiveness to science. In this
kind of relationship, personality syndromes can
function as mediators of the impact of family and
school on the development of attentiveness.

Second, students with different personality
dispositions will show different propensities to
participate in family and school activities that
foster attentiveness. Students with greater self-
esteem, a greater sense of the efficacy of individual action, or those who are less estranged from
other people may find it easier and more rewarding
to participate in the family and peer discussions
that are associated with the development of attentiveness to science. To the extent that these discussions foster the development of interest and knowledge of public issues, students with personalities
compatible with participation in discussions will be
likely to develop greater interest and knowledge of
the public issues that are discussed. In this kind
of relationship, family and school experiences can
function as mediators of the impact of personality
syndromes on the development of attentiveness.

A third alternative, of course, is that attentiveness to science and associated personality

traits arise together in certain family and school
environments and that the associations among these
common outcomes evidence no causal relationship.

Five family and school variables were identified in earlier chapters as having strong residual
associations with attentiveness: (1) gender, (2)
participation in family discussions of public issues, (3) participation in discussions of public
issues with peers, (4) the student's choice of occupation, and (5) the student's educational achievement and aspirations. Entering each of these five
social factors into the personality theory developed
in the current chapter raises the following questions:

1. Which personality factors appear more
   frequently in which kinds of social
   environments?

2. To what extent does the association of a
   social environment with a personality
   type suggest a causal relationship that
   helps us understand why some students
   become attentive to science and others
   do not?

The Role of Gender

The most fundamental social factor associated with
development of attentiveness to science and technology is the sex role -- or "gender" -- students have
played in their family and school environments. As
discussed in chapter nine, a large number of differences have been found in the ways parents, siblings,
and peers interact with boys and girls, and these
interactions can affect the development of attentiveness to science. The large, consistent differences in the attentiveness of men and women students
found in the present study are consistent with the
inference that these students have been treated in
systematically different ways regarding their development of attentiveness to science. It is interesting to ask, therefore, whether these differences in
treatment have produced any differences in the personalities these men and women have developed which
would make men more likely to become attentive to
science.

A path analysis tracing the impact of gender differences on the four personality factors and, subsequently, on attentiveness indicates significant personality differences between men and women students on three of the four personality traits (see Fig. 11.4). Women students in the NPAS sample are more likely than men to believe the informed citizen can be effective in changing public policies and are less likely than men to evidence estrangement from other people and their ideas. Men, however, are more likely than women to indicate high levels of self-esteem and self-confidence. Since two of these three differences would predict that women are more attentive to science than men, when the reverse is in fact the case, it is clear that the differences in personality between men and women do not account for their differences in tendencies to become attentive to science.

The residual association between gender and attentiveness, controlling for differences in personality, is by far the strongest association in the figure. Other factors that differentiate men and women are much more important than these personality differences in predicting their respective levels of attentiveness to science. This analysis indicates, then, that the socialization process develops personality characteristics, some of which are less compatible with the development of attentiveness in men than in women. Such differences must be overcome by other, more specific, influences if men tend to become more attentive to science than women in spite of these personality differences.

Family Politicization

The second social factor related to family environments is the frequency of family discussions of public issues. Chapter nine reported that students who discussed public issues more frequently with other members of their families were more likely to be attentive to science. Who are these students?

Frequency of family discussions is significantly associated with all four personality factors -- self-esteem and political efficacy, lack of estrangement, and trust in people (see Fig. 11.5). Students who talk more frequently with other members of their families about public issues tend to be higher in self-esteem, more optimistic about the

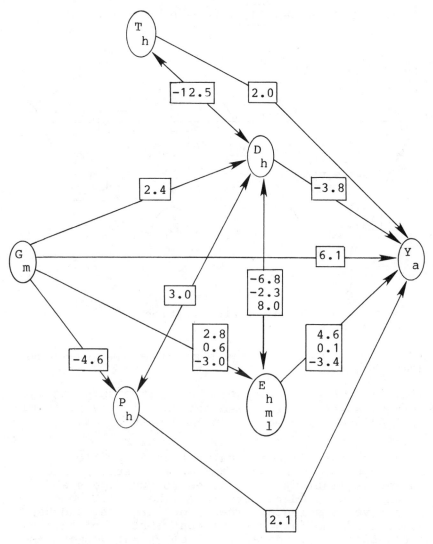

G = gender
  m   male

All other symbols are the same as in the preceding
table.

Fig. 11.4 A path model of the relationships among
attentiveness to organized science, gender, and
selected personality attributes.

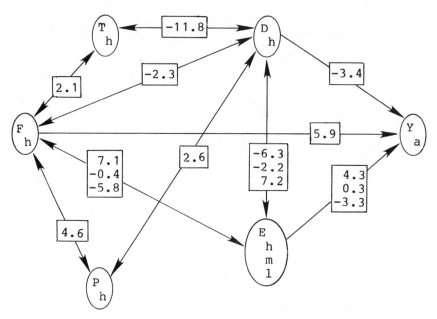

F = level of family politicization
h   high

All other symbols are the same as in the preceding table.

Fig. 11.5. A path model of the relationships among attentiveness to organized science, family politicization, and selected personality attributes.

effectiveness of individual action in public affairs, less estranged, and more trusting. Each of these associations contributes to our understanding of why these students are more likely to become attentive to science. These four factors together, however, do not exhaust the contribution of family discussion to the development of attentiveness, as evidenced by the residual association between family discussions and attentiveness controlling for all four personality variables.

Peer Politicization

A strikingly similar pattern of relationships emerges when relating personality factors with peer discussions, as we observed when comparing person-

ality factors with family discussions (see Fig.
11.6). Frequency of peer discussions is significantly associated with three of the personality
factors that were associated with family discussions
-- self-esteem, perceived efficacy of the informed
citizen, and estrangement. Students who are higher
in self-esteem and political efficacy and less
estranged from other people are more likely to talk
to peers about public issues. The residual association between peer discussions and attentiveness
again, however, indicates that other factors lead
students to become more attentive to these topics
whether or not they have personalities that are especially compatible with this development.

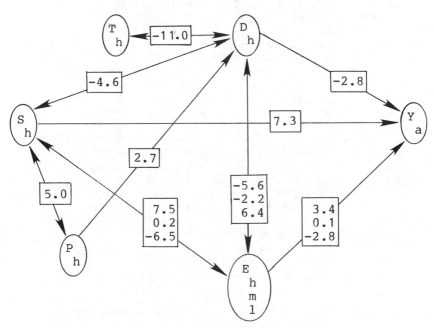

S = level of peer politicization
  h   high

All other symbols are the same as in the preceding
table.

Fig. 11.6. A path model of the relationships among
attentiveness to organized science, peer politicization, and selected personality attributes.

Occupational Preferences

Chapter ten reported that the kinds of occupations
students aspire to predict quite well to current
levels of attentiveness to science and technology.
Students who are planning to go into occupations
related either to science specifically or to public
service more generally are more likely to be attentive to science than are students who are planning
to go into other kinds of occupations. What kinds
of personality differences do we find among students
planning these kinds of occupations?
    The NPAS data indicate that differences in
occupational plans are associated with two personality factors -- self-esteem and political efficacy (see Fig. 11.7). Students who are planning a
career in science or public service tend to have
higher self-esteem than do students planning other
kinds of occupations and tend to perceive greater
efficacy in the individual actions. The residual
association between occupational choice and attentiveness, however, remains strong, controlling for
differences in these two personality factors. It
appears, then, that two of the reasons students
planning scientific and public service occupations
are more likely to be attentive to science are that
they tend to be the type of person for whom information about science and technology is useful in form-
ing their own opinions and influencing other people,
and they tend to perceive individual actions of
informed citizens as effective. Both these factors
would seem to provide incentives for students to
consider an occupation in the public arena and to
develop an attentiveness to public issues.

Educational Achievement and Aspirations

The strongest social factor predicting the development of attentiveness to science was educational
achievement and aspirations. Students who are
currently in high school and not planning to graduate from college are less likely to be attentive
to science than are their high school colleagues who
are planning to graduate from college, and high
school students as a whole are less likely to be
attentive to science than are college students. Are
there personality differences between high school

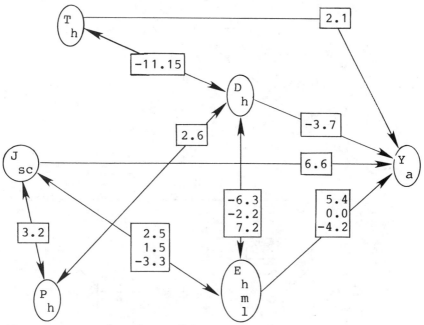

J   = occupational preference
sc   science

All other symbols are the same as in the preceding
table.

Fig. 11.7. A path model of the relationships among
attentiveness to organized science, occupational
preference, and selected personality attributes.

and college students that help us explain these
differences in level of attentiveness to science?
     Educational achievement and aspirations are
significantly associated with self-esteem, trust in
people, estrangement, and perceived efficacy (see
Fig. 11.8). College students and students planning
to attend college evidence higher self-esteem, greater trust in people, and greater efficacy than do
high school students who are not planning to attend
college. The major differences, however, are reflected in the students' levels of estrangement.
Students in college and students planning to obtain
a college degree are less likely to be estranged
from other people and their ideas than are high
school students not planning to graduate from
college.
     Taken together, the differences in educational
achievement and aspirations among people with dif-

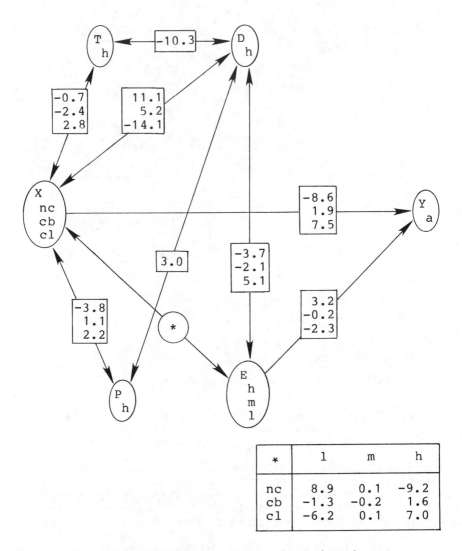

*	l	m	h
nc	8.9	0.1	-9.2
cb	-1.3	-0.2	1.6
cl	-6.2	0.1	7.0

X   = educational achievements and aspirations
  nc   HS-no coll
  cb   HS-coll bnd
  cl   college

All other symbols are the same as in the preceding table.

Fig. 11.8. A path model of the relationships among attentiveness to organized science, educational achievement and aspirations, and selected personality attributes.

ferent personalities suggest why some kinds of personalities appear more compatible with the development of attentiveness than others. Controlling for
educational achievement and aspirations, the only
personality factor that remains significantly associated with attentiveness is self-esteem. College
students and high school students planning to attend
college are more likely to have the kinds of personalities and skills that are compatible with the
development of attentiveness to science (and other
public issues), and once students sort themselves
out into those who are going to college and those
who are not, their individual personalities make
little difference in their becoming or not becoming
attentive to science. Whatever effects personality
differences have on the development of attentiveness, therefore, appear to operate in this sorting-
out process which determines who will and who will
not develop collegiate aspirations. The key to
understanding the relationship between personality
and attentiveness, therefore, seems to be in understanding the factors that contribute to the development of differences in students' perceptions of
themselves and other people. The role each of these
factors plays in determining whether a student decides to go to college seems to be much more important than any independent effects any of these
personality differences have in directly influencing
the development of attentiveness to science and
technology.

12 Attentiveness and General Political Interest

In the preceding three chapters, the roles of family, school, peers, and personality in the development of attentiveness to organized science were examined and a series of log-linear logit and path models were constructed. It is now appropriate to raise and examine an alternative hypothesis: that attentiveness to organized science is simply one manifestation of a general interest in political affairs rather than the issue specialization suggested in the preceding analysis.

To test this rival explanation, a measure of general political interest will be constructed and the three models proposed in the preceding chapters will be used to predict the new variable. If the models are as accurate in predicting general political interest as they were in predicting attentiveness to organized science, it will be necessary to conclude that attentiveness to organized science is, at least in part, a function of general political interest, negating the previous theoretical framework in large part. Conversely, if the models that predict attentiveness to organized science fail to provide an acceptable prediction of general political interest, it will be possible to conclude that attentiveness to science and technology issues is an analytically separate construct.

A MEASURE OF GENERAL POLITICAL INTEREST

The first task is the development of a measure of
general political interest. In the 1978 NPAS, several items reflected political interest and, following the multi-measure principle discussed earlier,
it is possible to combine several independent measures into a typology of general political interest.
     It is important to view political interest in
the context of competing family, career, and other
interests, since time and attention are essentially
zero-sum situations. In the 1978 NPAS, each respondent was asked to assess a set of 11 life goals and
to indicate whether each item was "essential," "very
important," "somewhat important," or "not important."
The 11 items were factor analyzed and three clear
factors emerged, which have been labeled family interest (FAMINT), career interest (CARINT), and political interest (POLINT). Although five items load on
the political interest factor, the first three items
are significantly stronger, thus the POLINT index
reflects the number of political interest items that
a respondent rated as "essential" or "very important." The index ranges from zero to three and is
distributed as follows:

Index	N	Percent
0	1718	42.7%
1	1207	30.0
2	679	16.8
3	425	10.5
Total	4029	100.0

     Political interest is weakly associated with
both family interest and career interest, but the
weakness of the relationship suggests the competitive nature of the various interest foci (see Table
12.1).
     While the POLINT index provides a good measure
of political interest as a life goal, it is possible
that some persons -- especially adolescents and
young adults just entering the political system --
may not identify politics as a major life goal, but
may still have a reasonably high level of interest
in electoral contests and other forms of political
competition. In recognition of this possibility,

the level of respondent partisanship was arrayed
against the POLINT index (see Table 12.2). An examination of the distribution of respondents on the
two measures suggests that both are useful measures
for some respondents; thus the final measure of
general political interest incorporates both partisanship and political interest as a life goal.

In the construction of a summary measure of the
salience of politics, all of the respondents scoring
two or more on the POLINT index were rated high on
political salience. This decision assumes that a
strong expression of interest in political events as
a major component of one's life mission is a good
indicator of probable political interest and participation. Similarly, all respondents reporting a
"strong partisanship" were classified as high on
political saliency. This decision reflects a judgment that an expression of strong partisanship in
the context of significantly lower levels of partisanship in most of the age cohorts included in the
1978 NPAS was a sufficient indication of political
salience.

While those two decisions handle the ends of
the distribution for the two measures, it is possible that a respondent might be a moderate in both
partisanship and commitment to politics as a life
goal, but the combination of the two factors would
still be sufficient to denote a positive interest in
political affairs. To capture this potential group,
those respondents with a score of at least one on
the POLINT index and who expressed a regular parti-

Table 12.1. Relationship of Family and
Career Interests

		Score on Career Interest Index						
Family Interest	N	0	1	2	3	4	5	Total
Low	754	5%	20%	18%	21%	20%	16%	100%
High	1451	3%	13%	20%	22%	21%	22%	101%

Gamma = .28

san affiliation were also scored as high on political salience.

The resulting typology produces a dichotomous
measure of political salience that includes approximately 39 percent of the respondents in a high political salience category and the remaining 61 percent
in a lower political salience classification. Compared to other measures of political interest or
saliency used by Jennings and Niemi (1974) and others, the total estimate of the proportion high on
saliency is similar. While various modifications or
refinements of the measure might have the effect of
moving the proportions by a few percentage points
one way or the other, it appears that the measure
just described reflects a reasonable division on salience that will provide a fair test of the alternative hypothesis outlined above.

Table 12.2. Relationship of Political Interests
            and Partisan Affiliation

Strength of Party Affiliation	N	Number of Political Interests				Total
		0	1	2	3	
Independent	1303	45%	29%	17%	9%	100%
Independent Party	641	36	34	18	12	100
Moderate Party	930	41	33	17	9	100
Strong Party	249	33	27	20	20	100
Missing	615	43	29	16	12	100

Gamma = .07 (based on 3123 cases, no missing value
cases included)

A COMPARISON OF THREE MODELS

Working from the measure of political salience outlined above and the attentiveness construct developed in chapters five, six, and seven, this section
of the analysis will compare each of three models in
terms of their ability to predict attentiveness to
organized science and political salience. The three
models will reflect the major components identified
in the preceding analyses of the impact of family,
school and peers, and personality.

In chapter nine, the analysis of the impact of
family and home environmental factors on the development of attentiveness to organized science identified four major sources of influence: parental SES,
gender, educational plans, and family politicization. When these four variables are entered into a
logit model, the four main effects (Model H2) produce a coefficient of multiple-partial determination
(Goodman 1972a, 1972b) of 0.92, which may be interpreted as explaining 92 percent of the total mutual
dependence in the model (see Table 12.3). Clearly,
the model is a very good predictor of attentiveness
to organized science.

When the same four variables are used to predict political salience, the coefficient of multiple-
partial determination drops to 0.76, a significant
decrease. Even though the model is significantly
less accurate in predicting salience than in predicting attentiveness, a model that explains 76 percent
of the available mutual dependence is not a terribly
bad fit either. It is appropriate, therefore, to
ask if the four variables in the model are equally
effective in predicting attentiveness and salience,
or if some of the variables are more useful in predicting one than the other.

An examination of the contributions of each of
the four main effects indicates that there are significant differences in the relative importance of
the independent variables in the two models. In the
prediction of attentiveness, educational plans is
the most important component of the model (see Model
H5), followed by gender and family politicization
(see Models H3 and H4). Parental SES is a relatively weak component of the model to predict attentiveness (see Model H6). In contrast, family
politicization is the major component in the prediction of salience, followed by educational plans.
Gender, which is the second most powerful predictor

Table 12.3. A Comparison of Family Effects Models

Models	When Y is attentiveness			When Y is salience		
	LRX^2	df	CMPD	LRX^2	df	CMPD
H1 GFTC,Y	541.53	23	--	161.09	23	--
H2 GFTC,YG,YF,YT,YC	42.57	18	.92	38.31	18	.76
H3 H2 minus YG	138.71	19	--	49.84	19	--
difference due to YG	96.13	1	.18	11.53	1	.07
H4 H2 minus YF	117.16	19	--	100.85	19	--
difference due to YF	74.59	1	.14	62.54	1	.39
H5 H2 minus YT	225.99	20	--	69.09	20	--
difference due to YT	183.42	2	.34	30.78	2	.19
H6 H2 minus YC	69.29	19	--	39.22	19	--
difference due to YC	26.71	1	.05	0.91	1	.01

G = gender   C = parental SES
F = family politicization   T = education plans

of attentiveness to organized science, is a weak
third in the model to predict salience. Social
class (parental SES) is not a significant component
of the model to predict salience, although it was a
significant (but weak) part of the attentiveness
model.

Looking at the fit of the variables in the family effects model in predicting attentiveness and
salience, it is apparent that the model that pre-
dicts attentiveness is not a good predictor of salience and that the relative role of the components
of the family effects model in the two predictive
tests is significantly different.

In chapter ten, the analysis of school and peer
effects identified three additional components that
were useful in predicting attentiveness: occupational preference, peer politicization, and the
number of different scientific disciplines studied.
To provide for an examination of school and peer
effects in the context of the family effects, which
logically and temporally preceded the school experience, family politicization, gender, and educational
plans were carried forward, producing a model with
six independent variables. Using the log-linear
logit model described earlier, the main effects
model produces a coefficient of multiple-partial
determination of 0.84, which is a very good fit (see
Model H2 in Table 12.4).

When the same six-variable model is used to
predict political salience, the coefficient of
multiple-partial determination drops to 0.56, which
is not a very tight fit. In short, approximately 44
percent of the total mutual dependence in the model
remains unexplained. Further, looking at the relative contributions of each of the independent variables in the two models, a number of significant
differences appear. Gender continues to be a valuable component of the model to predict attentiveness
to organized science and a very weak contributor to
the prediction of political salience. Both family
politicization and peer politicization are better
predictors of political salience than of attentiveness.

In summary, a comparison of the combined
school-family effects model in the prediction of
attentiveness and salience indicates that the attentiveness model is not an adequate predictor of
political salience. In terms of the original alternative hypothesis, this result would suggest that

Table 12.4. A Comparison of School and Peer Effects Models

	Models	When Y is attentiveness			When Y is salience		
		LRX^2	df	CMPD	LRX^2	df	CMPD
H1	FGOTPS,Y	636.82	95	--	286.30	95	--
H2	FGOTPSY,YF,YG,YT,YO,YP,YS	102.41	88	.84	124.85	88	.56
H3	H2 minus YF difference due to YF	132.85 30.44	89 1	-- .05	151.06 26.21	89 1	-- .09
H4	H2 minus YG difference due to YG	172.42 70.01	89 1	-- .11	132.45 7.60	89 1	-- .03
H5	H2 minus YT difference due to YT	225.20 122.79	90 2	-- .19	152.60 27.75	90 2	-- .10
H6	H2 minus YO difference due to YO	137.24 34.83	89 1	-- .05	134.79 9.94	89 1	-- .03
H7	H2 minus YP difference due to YP	136.94 34.53	89 1	-- .05	150.25 35.40	89 1	-- .09
H8	H2 minus YS difference due to YS	106.15 3.73	89 1	-- .01	132.74 7.89	89 1	-- .03

F = family politicization   O = occupational preference
G = gender   P = peer politicization
T = educational plans   S = number of science areas studied

the attentiveness construct is significantly different from the political salience construct.

In chapter eleven, the analysis of the role of
personality variables in the development of attentiveness to organized science identified two additional variables that should be added to an overall
model: self-esteem and social estrangement. As
above, these variables were examined in the context
of the previous family-school model, resulting in a
model with the following independent variables:
family politicization, gender, educational plans,
self-esteem, social estrangement, and parental SES.

In the prediction of attentiveness, the main
effects model produces a coefficient of multiple-
partial determination of 0.74, accounting for about
three-quarters of the total mutual dependence in the
model (see Model H2 in Table 12.5). In general,
this is an acceptable fit, although not as good a
predictor as the earlier models. It should be noted
that each of the three models examined have added
increasing levels of complexity; thus the coefficient of multiple-partial determination has dropped
somewhat. On the grounds of accuracy and parsimony,
it is not clear that the latter, more complex models
are major improvements over the first and simpler
model, but the purpose of the latter models was to
examine the role of various new independent variables in a comparative context, and for that purpose, the additional models have been most useful.

In contrast to the prediction of attentiveness,
the main-effects model that includes the personality
variables produces a coefficient of multiple-partial
determination of only 0.19, a clearly unacceptable
level of prediction. Interestingly, the two personality variables turn out to be slightly better predictors of general political salience than they do
of attentiveness, viewed in the context of the
family and school effects (see Table 12.5).

In summary, while the personality variables are
indeed better predictors of political salience than
attentiveness, the total model remains a significantly better predictor of attentiveness than of
salience. The observation that the two personality
variables are better predictors of salience than
attentiveness is a useful suggestion of the role
that they may play in the attentiveness model, that
is, providing a tie to general political salience.

Table 12.5. A Comparison of Personality Effects Models

Models	When Y is attentiveness LRX^2	df	CMPD	When Y is salience LRX^2	df	CMPD
H1 FGETCS,Y	637.54	143	--	249.76	107	--
H2 FGETCS,YF,YG,YE,YT,YC,YS	167.60	135	.74	201.49	135	.19
H3 H2 minus YF	236.01	136	--	257.98	136	--
difference due to YF	68.41	1	.11	56.49	1	.23
H4 H2 minus YG	259.83	136	--	211.25	136	--
difference due to YG	92.23	1	.14	9.76	1	.04
H5 H2 minus YT	284.72	137	--	224.80	137	--
difference due to YT	117.12	2	.18	23.31	2	.09
H6 H2 minus YE	169.65	136	--	210.27	136	--
difference due to YE	2.05	1	.00	8.78	1	.04
H7 H2 minus YC	196.41	136	--	202.20	136	--
difference due to YC	28.81	1	.05	0.71	1	.00
H8 H2 minus YS	178.09	137	--	214.61	137	--
difference due to YS	10.49	2	.02	13.12	2	.05

G = gender
F = family politicization
T = educational plans

E = social estrangement
C = parental SES
S = self-esteem

SUMMARY

Returning to the alternative hypothesis outlined at the beginning of this chapter, it is now possible to assert that the models that are effective predictors of attentiveness to organized science are significantly less effective predictors of political salience. While the results would suggest that the two constructs -- attentiveness and political salience -- are not unrelated, there are sufficient differences between the two typologies so that it is analytically and conceptually useful to treat them separately.

13 A Developmental Model

On the basis of the analyses reported in the preceding chapters, it is now possible to identify a final set of independent variables that are useful in understanding the development of attentiveness to organized science by young Americans. The rationale for the selection of each of the final variables will be reviewed and then a logit and a path model will be presented.

THE INDEPENDENT VARIABLES

The final model includes five independent variables: occupational preference, educational plans, politicization, gender, and self-esteem. Each of the variables is a significant and contributing component of the final model at the 0.01 level, although some account for a substantially larger proportion of mutual dependence than others. Each of the variables represents an important conceptual facet of the developmental process, and it is useful to review briefly the reasons for including each of the variables in the final model.

Gender is logically prior to all of the other variables in the model and explains a significant amount of mutual dependence, even holding constant all of the other variables in the model. In substantive terms, it appears that young women are significantly less likely to be interested in and informed about organized science issues than young men, even taking into account differences in educational and

career plans. In chapter ten, it will be recalled, gender persisted as a significant component
even when the number of science areas studied and
other school and peer variables were held constant.
Given the persistence of the gender differences, it
is necessary to conclude that the early socialization of young women appears to create a lack of
interest in science and technology issues.

The educational-plans variable, also referred
to as the school typology, classifies the respondents as high school college-bound, high school
non-college-bound, and college students. This distinction splits the total study sample into almost
equal thirds and has proven to be one of the most
powerful predictors of attentiveness to organized
science of all of the variables utilized in the previous models. Those high school students not planning to attend college appear to be almost oblivious
to science and technology issues, expressing neither
interest in nor information about any of the topics
in that domain. In contrast, the college-bound high
school student is significantly more interested in
science and technology issues and substantially better informed. And the transition to college appears
to stimulate even higher levels of interest in and
knowledge about organized science. These patterns
would suggest that the decision to be attentive to
science and technology issues is probably closely related to occupational and professional plans and is,
perhaps, utilitarian in nature. By the same token,
it could be argued that college-bound high school
students are usually exposed to more and better science courses than non-college-bound students and
that greater knowledge about science generates interest and stimulates the retention of information
essential to issue awareness. Regardless of the
path or motive, it is clear that the educational
plans variable is strongly associated with attentiveness to organized science and that it is a major
predictor of attentiveness in every model tested in
this research.

The occupational-preference variable follows
the same logic and is useful in seeking to separate
those persons interested in a career in a science-
related area and those who are attentive even though
they do not have a career interest in a science- or
technology-related field. This variable tended to
be somewhat weaker than educational plans in its
association with attentiveness in the multivariate
models in the previous chapter, but it is concep-

tually important in that it allows a distinction
between those young people anticipating a science or
public service career and those who expect to pursue
other occupations and professions.

The politicization variable in the final model
is a combination of the family-politicization and
peer-politicization variables used in previous models. In examining the interaction between family
and peer politicization, it became apparent that
family politicization was dominant when it existed
at all and that peer politicization was effective
primarily in the instance of low or nonexistent family politicization. Interestingly, when a respondent
experienced neither peer nor family politicization,
the person tended to adopt attitudes even closer to
his or her family than did respondents who experienced a high level of family politicization. These
observations, led to the creation of a three-fold
politicization variable that would reflect strong
family politicization, peer politicization in the
absence of family politicization, and the absence of
both peer and family politicization. Since both
family and peer politicization were good predictors
of attentiveness separately, it is anticipated that
the combined variable will be an even better predictor of attentiveness to organized science.

Finally, the self-esteem variable was retained
in the final model for its indirect contributions to
attentiveness through educational plans and politicization. In view of the significant, but weak,
contribution of self-esteem in the models in chapter
twelve, it is anticipated that this variable will
have a weak main effect and will be most useful in
the path analysis of the indirect influences on the
growth of attentiveness.

Having reviewed the five independent variables
included in the final model, it is useful to explain
briefly the reasons for not including some of the
other variables that appeared in some of the models
in previous chapters. Parental SES, or social class,
is an excellent predictor of educational plans, but
in the multivariate models examined earlier in the
analysis, it appears that there are few direct, or
main, effects associated with social class, but that
its contribution is indirect and expressed almost totally through the educational-plans variable. Since
the educational-plans variable is included in the
final model, it would be redundant to include the
social class measure.

Several of the educational variables were considered for inclusion, but they were relatively weak
contributors in the multivariate models, suggesting
that their influence was expressed through other
variables like educational plans and occupational
preferences. When variables like the number of scientific disciplines studied and class rank were examined, they were very weak contributors to the
total model when educational plans and occupational
preference were held constant. Accordingly, only
occupational preference and educational plans were
incorporated into the final model.

Social estrangement was a relatively strong component in the personality model, but in the larger
multivariate models that included the school and
family effects, it was relatively weak and appeared
to be associated only with self-esteem. Since self-
esteem is included in the final model, the marginal
contribution of the social estrangement measure appeared to be inadequate to justify its inclusion in
the final set of variables.

A LOGIT MODEL

The first step in the analysis of the final model is
the construction of a log-linear logit model, following the procedures outlined by Goodman (1972a,
1972b). The base model (Model H1) indicates that
there are 655.21 chi-square of mutual dependence in
the total model, and the main effects model (Model
H2) accounts for all but 123.0 chi-square of mutual
dependence (see Table 13.1). This model produces a
coefficient of multiple-partial determination (CMPD)
of 0.81, which may be interpreted as explaining 81
percent of the total mutual dependence in the model.
In short, the main-effects model provides an excellent fit for the prediction of attentiveness to organized science. None of the higher order effects
were significant at the 0.01 level.

Looking at the contribution of each of the components in the model, the largest proportion of mutual dependence is accounted for by a respondent's
educational plans or current educational status in
the case of college students. This single variable
accounts for 17 percent of the total mutual dependence in the model (see Model H7).

The second best predictor in the model is the
politicization variable described above. This vari-

able accounts for about 15 percent of the total mutual dependence (see Model H5). It appears that the
combination of the family-politicization and the
peer-politicization variables into a single measure
produced a significantly stronger variable than
either of the individual measures.

Gender is the third most significant component
in the model, accounting for about 11 percent of the
total mutual dependence in the model (see Model H6).
This result is consistent with the role of gender in
several of the models discussed in previous chapters
and demonstrates the long-term effects of early socialization about subjects that are "appropriate"
for males and females.

Table 13.1. The Main Effects Model to
Predict Attentiveness

Models	LRX^2	df	CMPD
H2 OSPGY,YO,YS,YP,YG,Yt	123.00	99	.81
H3 H2 minus YO	162.78	100	--
difference due to YO	39.78	1	.06
H4 H2 minus YS	137.21	100	--
difference due to YS	14.20	2	.02
H5 H2 minus YP	221.45	101	--
difference due to YP	98.44	2	.15
H6 H2 minus YG	195.06	100	--
difference due to YG	72.06	1	.11
H7 H2 minus Yt	235.43	101	--
difference due to Yt	112.43	2	.17

Y = attentiveness to organized science
O = occupational preference
S = self-esteem
G = gender
t = educational plans
P = politicization

The fourth largest contributor to explain mutual dependence is the occupation-preference variable, which accounts for about 6 percent of the total mutual dependence in the model (see Model H3). As noted above, it was assumed that this variable would co-vary with educational plans and that the strength of the educational-plans variable would have the effect of reducing the main effect of the occupational preference measure by itself. Nevertheless, it is a significant contributor to the total model.

The least important component of the model in terms of its direct effect of the development of attentiveness to organized science is the self-esteem variable. Its main effect accounts for only 2 percent of the total mutual dependence (see Model H4).

The logit model is designed to provide an assessment of the direct and relative effect of a series of independent variables on a dependent variable, in a manner analogous to a regression equation. The data in Table 13.1 provide a useful set of relative measures and suggest several important interpretations. Before turning to a substantive discussion of the implications of the model, it is useful to examine a path model utilizing the same variables. This examination leads to a better understanding of the relationships among several of the independent variables.

A PATH MODEL

While path models have been traditionally used with interval level data, Goodman (1972a, 1972b) has suggested a method for utilizing the log-linear model approach to construct causal models analogous to the traditional path analysis. In the analysis below, the procedures outlined by both Goodman and Fienberg (1978) have been employed. The path coefficients reported in the model are standardized values, which are produced by dividing either the lambda or beta coefficients by their respective standard errors. A standardized value of 1.96 or more is significant at the 0.05 level.

The complete path model for the five independent variables and attentiveness is displayed in Figure 13.1. The arrows and path coefficients have been included for all relationships in which at least one of the coefficients was significant at the 0.05 level. The absence of an arrow between two

variables means that none of the standardized values
were significant. For each of the variables in the
model, the subscript denotes the value or values for
which coeffients are reported.
    For the reader unfamiliar with log-linear path
models, it may be useful to review some of the major
paths and to interpret the standardized values.
Starting with gender, the model indicates that gender has a direct effect on attentiveness and several
indirect effects, all of which work through the politicization variable. Looking at the direct effect,
the standardized value of -4.4 means that women are
significantly less likely to be attentive to organized science, holding constant all of the other
independent variables. A value of 4.4 is relatively
substantial, looking at the other standardized values in the model. The arrow between P and G means
that the level of politicization differs significantly according to gender. The three subscripts of
P represent family politicization (F), peer politicization in the absence of family politicization
(P), and the absence of both peer and family politicization (N). The standardized values indicate that
the gender differences in the level of family politicization are not significant at the 0.05 level,
suggesting that in this study families are not more
likely to discuss politics with male children than
female children. The standardized value of -2.6
indicates that young women are significantly less
likely to discuss political topics with their peers
than are young men. The standardized value of 2.5
means that young women are significantly more likely
to discuss political subjects with neither family
nor peers than are young men. Since family and peer
politicization are positively related to attentiveness to organized science, it would appear that
young women are less likely to engage in peer political discussions than young men and that the
absence of this discussion exposure or experience
results in a lower level of issue interest and
awareness by young women.
    The absence of an arrow can be as significant
as the presence of a relationship, and it is interesting to note some of the relationships in which no
significant gender difference appeared. The absence
of a significant GS relationship means that young
men and young women do not differ in self-esteem.
The absence of a significant GO relationship means
that young women are as likely as young men to aspire to a career in a field related to science or

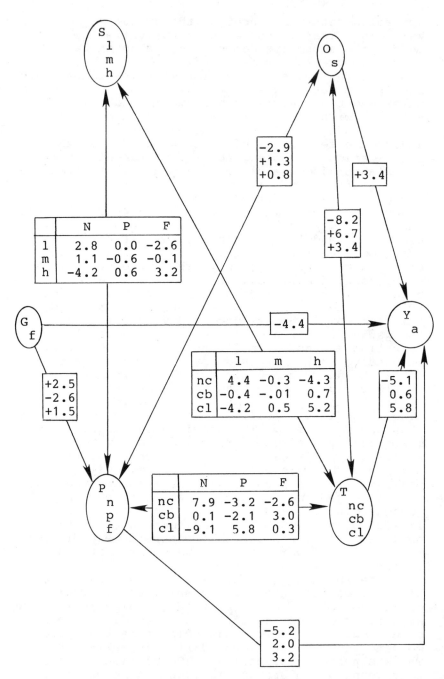

Fig. 13.1.   The complete path model.

public service. The absence of a GT relationship
means that young women are as likely to go to college as young men. In view of the absence of significant differences in self-esteem, occupation preference, and educational plans, it is particularly
surprising to find the strong negative direct effect
of femaleness on the level of attentiveness to organized science. The difference would appear to lie
in the politicization variable, and it is reasonable
to suggest that young women do not view politics as
either an appropriate or interesting subject for
discussion, especially among their peers.
      In view of the complexity of the complete path
model and the interactions between several of the
polytomous variables, it is useful to seek to identify some of the major paths in the complete model
and to examine those paths separately. An examination of several possible paths suggests a simplified
model for the prediction of attentiveness to organized science and a parallel model of the prediction
of nonattentiveness.
      Looking first at the model to predict atten-
tiveness to organized science (see Fig. 13.2), the
indirect effect of self-esteem becomes apparent.
Those respondents who were high in self-esteem were
significantly more likely to be in college and to
participate in political discussions with other family members than were persons with lesser levels of
self-esteem. While self-esteem does not have a significant main effect on attentiveness to organized
science, the encouragement of family politicization
and college attendance are important avenues to the

Legend for Fig. 13.1.

G	= gender		S	= level of self-esteem
f	female		l	low
			m	medium
P	= type of politicization		h	high
n	none			
p	peer only		T	= educational plans
f	family, with or		nc	HS-no coll
	without peer		cb	HS-coll bnd
			cl	college
O	= occupational			
	preference			
s	science			

Fig.13.1.   The complete path model.

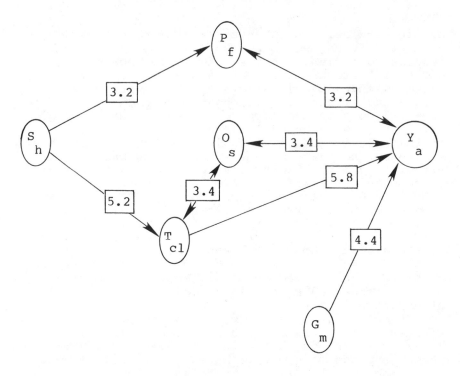

All symbols are subsets of the variables used
in Fig. 13.1.

Fig. 13.2.   A simplified path model to predict
attentiveness to organized science.

subsequent development of attentiveness. College
attendance has both a direct effect on attentiveness
(S.V. = 5.8) and an indirect effect through occupational preference (S.V. = 3.4).   Occupational preference, in turn, has a direct effect on attentiveness
(S.V. = 3.4).   Family politicization has a similar
direct effect on attentiveness (S.V. = 3.2).
     In this simplified model, gender is unrelated
to any of the other independent variables, but has a
strong direct effect (S.V. = 4.4) on attentiveness,
positive for men and negative for women.   It is
important to set this effect in the context of the
previous path.   The complete model indicated that a
young woman should have about the same probability
of going to college, discussing politics with her
family, or seeking a scientific or public service
career as a young man, but her gender (perhaps

through earlier socialization) appears to be a negative offset to the otherwise positive probabilities
of attentiveness.

An examination of the simplified model for the
prediction of nonattentiveness will provide some additional insights into this process (see Fig. 13.3).
The indirect role of self-esteem is again apparent,
with those respondents scoring low on self-esteem
being significantly more likely to not attend college (S.V. = 4.4) and to have no political discussions (S.V. = 2.8) than their colleagues with higher
levels of self-esteem. Each of these choices leads
to both direct and indirect paths that are negatively associated with attentiveness. Non-college attendance has a significant negative direct effect on
attentiveness (S.V. = -5.1) and a significant negative indirect effect through occupational preference

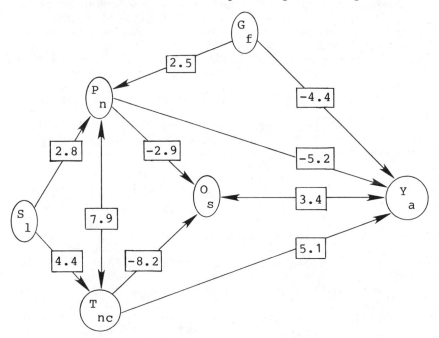

All symbols are subsets of the variables used
in Fig. 13.1.

Fig. 13.3. A simplified path model to predict nonattentiveness to organized science.

(S.V. = -8.2). Similarly, the absence of politicization is directly negatively associated with
attentiveness (S.V. = -5.2), and it is indirectly
negatively associated with attentiveness through
occupational preference (S.V. = -2.9). In short,
this model indicates that respondents with low self-
esteem are unlikely to attend college or engage in
political discussions, and that persons who do not
attend college or discuss politics are not likely to
seek careers in scientific or public service fields
or to follow issues involving science and technology.
     The role of gender in this model is important.
Gender has a direct negative effect on attentiveness
for women (S.V. = -4.4) and is also positively associated with the absence of political discussion
(S.V. = 2.5), which, in turn, is strongly and negatively related to attentiveness. This combination
of paths suggests that young women are somewhat more
likely than young men to have no political discussions at all (even though they are equally likely to
talk to their own families about political issues),
and this absence of politicization is a strong negative factor in the development of attentiveness.
Given that a woman's gender alone already has a
strong negative impact apart from the other independent variables, the absence of politicization may be
seen as an additive effect, in contrast to the off-
setting relationship in the earlier model to predict
attentiveness.

                         SUMMARY

Examination of the final model in both its logit and
its path forms suggests that exposure to college
education is the single most important factor in the
development of attentiveness to organized science.
Both family politicization and occupational preference for a scientific or public service career are
also important factors in the emergence of attentiveness in young adults. The final model indicates
that self-esteem plays an indirect, but important,
role in both family politicization and educational
plans. While the level of self-esteem did not differ significantly by gender, the models revealed
that gender sometimes combined with low self-esteem
to increase the probability of low levels of politicization, which in turn reduces the probability of
being attentive to organized science.

V
The Implications of Attentiveness

14 Citizenship in an Age of Science

At this point, it is necessary to return to two basic questions that were raised briefly in the early discussions: What is the role of attentive publics in a participatory political system like that of the United States? What is the significance of the preceding analysis concerning the attentive public for organized science for the American political system? These are important questions and deserve careful consideration.

THE ROLE OF ATTENTIVE PUBLICS

In his original work, Almond (1950) realized the tension between the existence of smaller specialized publics and democratic principles. Was it acceptable in a democracy to allow a very small minority (even if they were highly educated and articulate) to shape an important area of public policy like foreign affairs? The nature of mass society, Almond concluded, made it impossible for all citizens to be equally aware of all policy areas, but it was essential that the general public retain a final veto authority should they be motivated to apply it. The public reaction to the Korean War in the early 1950s and to the Vietnam War in the 1970s demonstrated the ability of the citizenry to react to unpopular foreign policies and to assert the influence of their numbers through the electoral process.

The idea that it is impossible for all citizens to be actively knowledgeable in all areas of public

289

policy was further sharpened by Dahl's concept of
polyarchy. Looking at the full range of systems of
governance in political systems, Dahl (1971) describes the two ends of the spectrum as autocracy
and democracy. Much of classical political theory
contrasts these pure types, but Dahl argues that in
reality political systems tend to fall along a spectrum between the two extremes. He asserts that the
inherent inequality of citizens in regard to resources and ability means that it is impossible to
have a genuine democracy in any populous political
system. The mixed form of government that tends
toward democracy but does not achieve the optimal is
termed a polyarchy.
     Among the feasible choices for political sys-
tems, Dahl and numerous other political theorists
argue that a polyarchy is the most desirable in that
it provides an ultimate popular veto power on the
actions of government, but allows those groups of
citizens with specialized interests and abilities to
debate and initiate policy proposals in a wide variety of substantive areas. This line of argument has
been extended by Downs (1957), Tullock (1972), and
other political scientists who have argued that a
person's political behavior can be understood best
via an economic model of man. In this view, each
person has a certain amount of time and other scarce
resources and must allocate these resources so as to
maximize his or her income, happiness, or any other
goal he or she may cherish. This means that persons
with little education in science and little professional or political interest in the area would be
expected to be reluctant to devote large amounts of
time and other resources to becoming informed at a
high level about the issues and alternatives in
science policy. In short, mass political systems
cannot achieve the ideal of democracy, and it is
necessary that smaller and specialized segments of
the public provide policy leadership in these are-
nas, and both foreign policy and science policy are
prominent examples.

                    The Question of Size

If it is accepted that mass political participation
is not feasible on every set of issues active in a
political system and that a certain amount of policy
leadership must be provided by elites and attentive

publics, it is appropriate to ask how large attentive publics should be, or if there is an optimal
size. More specifically, it is reasonable to ask if
the probable size of the attentive public for organized science is sufficient. Extrapolated into the
next decade from the 1978 NPAS, it would be about 20
percent of the adult population.

The answer to these questions involves two factors -- accessibility and a time perspective. The
issue of access is central. Are there barriers to
becoming attentive to a set of issues? Or, in more
economic terms, is the price of access reasonable?
By and large, in the American political system, membership in an attentive public is as informal and
open as membership in a political party. American
political parties have not and do not require membership applications or dues, or set other requirements.
In fact, under the voting laws of most states, a
citizen may change his or her party allegiance free-
ly and easily and typically can vote in either
party's primary (but in only one).

In a parallel manner, attentiveness is a self-
selected attribute -- that is, an individual decides
that he or she is sufficiently interested in science
policy, or foreign policy, or civil rights to make
the effort necessary to keep informed about the issues in that arena. It is not a formal decision,
but rather is reflected in a broad pattern of personal behaviors -- reading selections, discussions
with friends and family, listening to specialized
shows on radio or television, and similar actions
designed to learn what is happening on some topic.
For those persons who are sufficiently interested,
informed, and aroused about an issue or cluster of
issues, the behaviors may extend to joining one or
more formal organizations, attending meetings, organizing rallies or other actions, recruiting candidates, writing to officeholders, and a variety of
other gladiatorial political activities. All of
these behaviors, however, take time and effort, and
it is the series of day-to-day decisions about these
various forms of behavior that determines attentiveness, not any central group or entity. In this
sense, most attentive publics are highly accessible
to interested citizens.

The question of size must also be set in a time
context. Had a study similar to the 1978 NPAS been
conducted in 1968, it is likely that the level of
interest in foreign policy among young adults would
have been substantially higher than any other polit-

292 CITIZENSHIP IN AN AGE OF SCIENCE

ical issue. In fact, a series of questions by the
Gallup Organization to the adult population of the
United States about the most important issues facing
the United States showed a major increase in foreign
policy issues during the peak of the Vietnam involvement, and a decline in foreign policy interests
during the years after the end of the American involvement in the fighting.

The Vietnam example is particularly useful in
addressing the question of an optimal size. As the
1978 NPAS data demonstrate, foreign affairs are traditionally a low saliency topic for most Americans,
with only about 10 percent of the adult population
qualifying as attentive (Almond 1950; Rosenau 1961,
1963, 1974). Yet, when the government's actions in
Vietnam became increasingly unpopular, more and more
individuals who two years earlier could not have
spelled Vietnam became conversant with the DMZ,
DaNang, American tactical policies, kill ratios,
and the internal politics of the Saigon government.
The broadcast and print media provided an increasing
flow of information; candidates began to include
their positions on the war in their public statements; and the war issue began to push numerous
other issues to the back burner of politics. Once
the American involvement stopped and U.S. troops
were withdrawn, attentiveness to foreign policy declined sharply and policy leadership returned once
more to the elites and attentives that dominate in
normal times. The essential point is that within a
period of less than four years, the attentive public
for foreign policy swelled to mass proportions, successfully reversed the government's war policy, and
then receded to its normal lower level. Accessibility to an attentive public is clearly open in the
American political system.

From these observations, it is reasonable to
conclude that the primary question is accessibility,
not size, and that size is not a stable characteristic, but rather susceptible to change if the mass
public rejects the policies or programs of the incumbent government. In the case of science issues,
the recent accident at Three Mile Island apparently
succeeded in increasing public awareness of the nuclear power controversy, although it is far from
clear that this brief peak of interest can be sustained without further major accidents of the same
character.

Are Attentive Publics Elitist?

Any time that public policy leadership is taken by a
small group of citizens, it is important to ask if
this represents an elitist movement, or if it ex-
presses a pluralism in the system. While the literature concerning the elitist-pluralism argument about
the American political system is vast and beyond the
scope of this monograph, it is useful to apply a
simple test to the 1978 NPAS data to determine the
structure of young adult participation in the political system in regard to issue attentiveness. In
general terms, if the same group of respondents turn
out to be attentive to organized science, foreign
policy, economic policy, and civil rights issues, or
any significant clustering of three areas, then it
would be necessary to conclude that this small group
would qualify as a future elite, exerting policy
leadership in several important areas. On the other
hand, if the pattern of attentiveness is diverse,
with few respondents having more than two issue foci,
then it is reasonable to term this pattern pluralist,
since there would be little overlap in the policy
leadership groups in these issue areas.
     An examination of the distribution of atten-
tiveness by the school typology indicates that attentiveness to issues is primarily an activity of the
college educated in the American political system.
Almost 90 percent of the non-college-bound high
school students (who represent approximately half of
their respective age cohorts) are attentive to none
of the four topic areas covered by the 1978 NPAS
(see Table 14.1). While it is possible that some
persons in this group are attentive to issues other
than the four in the NPAS, it is not likely that a
significant number of people would be attracted to
other interests, since the topics included in the
NPAS tend to be those most often cited in Gallup's
surveys of adult interests. Even if the nonattentiveness figure were 80 percent, the basic point is
that most non college-educated Americans have no
issue specialization and are therefore limited in
their political participation to electoral activities and to broad group allegiances.
     In contrast, approximately 56 percent of the
college population qualified as attentive to one or
more of the four issue areas covered by the 1978
NPAS. Five percent of the college population were
attentive to all four issue areas, and an additional

Table 14.1. The Distribution of Attentiveness
            to Four Political Issue Areas

Subject of Attentiveness	HS-No Coll	HS-Coll Bnd	College	Total*
SFEC	.1	1.5	4.9	2.2
SFE	.2	.9	1.8	1.0
SFC	.2	.7	1.8	.9
SEC	.7	1.8	4.7	2.4
FEC	.2	.6	4.4	1.9
SF	.4	1.2	1.5	1.0
SE	.4	1.9	2.5	1.6
SC	.5	2.3	4.8	2.5
FE	.1	.8	.8	.6
FC	.0	.6	1.6	.8
EC	.9	2.6	6.8	3.4
S	2.1	6.6	7.2	5.2
F	.5	1.7	1.4	1.1
E	2.9	4.7	5.8	4.4
C	1.4	5.6	5.8	4.2
None	89.4	66.4	44.1	66.9
Total	100.0	99.9	99.9	100.1
N =	1217	1098	1421	4028

* Total includes 292 high school respondents who
   did not report their college plans.

S = attentive to organized science issues
F = attentive to foreign affairs issues
E = attentive to economic issues
C = attentive to civil rights issues

12 percent were attentive to a cluster of three issue areas. A full 18 percent of the college population were attentive to a pair of issues, and 20
percent were attentive to a single issue. While the
distribution of issue interests within the college
population is pluralist in nature, the contrast of
the college and non-college populations suggests
that attentiveness is strongly associated with higher
education in all four issue domains, thus implying a
significantly different political role for the better educated citizens versus the less well-educated
citizens. To some extent, this result is probably
inherent in the concept of polyarchy, since advanced
education and the skills derived from that experience are among the chief inequalities inherent in
mass political systems, according to Dahl.
     The answer to the pluralism-elitism question,
then, is a mixed one. There is not a single small
group that dominates policy leadership in the four
policy areas studied, but there is a strong association between attentiveness and a college education.
As larger and larger proportions of the adult population become college educated, this inequality may
diminish, although it is possible that the educational threshold will increase, thus continuing the same
type of division observed currently. And there is
some evidence that this latter alternative may occur. Although the school typology utilized in this
analysis grouped all college students together, a
number of preliminary analyses not reported earlier
indicated that college students planning to attend
graduate or professional school were about 10 percent more attentive to organized science than those
not planning to continue beyond the baccalaureate.

Attentive Publics and Electoral Politics

Since the single most frequently investigated set of
questions in American politics concerns the roots of
and reasons for voting behavior, it is useful to
focus briefly on the relationship between attentive
publics and the traditional electoral processes in
the American polyarchy.
     In normal times, almost all of the efforts of
the elites and their respective attentive publics
are conducted outside the electoral process. The
attentive public for organized science is an excellent example. The basic objectives of the American

scientific community, its elite leaders, and a substantial majority of the attentive public for organized science may be summarized as: (1) a continued
high level of public financial support for both basic and applied research, but especially the former;
and (2) a minimum of external regulation or intervention in the activities of the scientific community. These goals have not changed substantially
since they were first and forcefully enunciated by
Vannevar Bush in 1946. To a very large extent,
these goals have been and continue to be achieved.

To effect these outcomes, the leadership of the
scientific community (the elites, in Almond's terms)
have been able to establish and sustain excellent
communication channels with the members and staff of
the relevant congressional committees and executive
agencies. They bring to the committees a corps of
experts who are ready to provide testimony on key
issues and to assist members and staff in understanding the issues in less technical terms. The
American Association for the Advancement of Science
(AAAS) is the principal coordinating group for the
scientific community, although numerous private corporations active in scientific and technological
affairs and many professional and disciplinary societies make separate presentations on various issues.
In this case, the technical nature of many of the
issues discourages intervention by groups lacking a
high level of expertise, although there are an increasing number of antiestablishment groups within
the scientific community, many of which do provide
some expert testimony, but few of which can match
the AAAS for an ongoing legislative affairs program.

Similarly, to prevent unwanted regulation or
intervention, the scientific community works through
the AAAS and through its disciplinary professional
societies to seek to influence both legislators and
the administrators of executive agencies to support
peer review and self-regulation from within the scientific community. While many federal agencies draw
expertise from a wide range of groups and institutions, often without any consultation or review by
the leadership organizations of the scientific community, the AAAS and the professional societies are
able to place significant numbers of their own leaders in key advisory positions. This descriptive
review is not meant to imply that this process of
informal influence is not legitimate, for these organizations and groups include some of the most
knowledgeable and eloquent spokespersons available

from the scientific community on the principal issues at any point in time. By and large, the scientific community has been true to its commitment to
open discussion, and there is no evidence of a con-
certed effort by the established groups to silence
dissent on any of the major issues before the Congress or the executive agencies. Yet, inevitably,
dissident voices lack the mantle and the aura of the
established leadership and generally are less successful in achieving their policy objectives than
are the AAAS and the major groups.
     The AAAS and all of the major scientific so-
cieties are strongly committed to a nonpartisan
position and do not endorse or oppose particular
candidates for public office. They do not contribute to campaigns or organizations, and few of their
leaders engage in active electoral work on behalf of
any candidates. To a very large extent, the relatively small size of the scientific community would
make direct electoral intervention foolhardy. Further, the provision of technical expertise to the
winners from all parties has been so successful up
to this point that there would appear to be little
reason to search for a different strategy.
     Like the scientific community and its attentive
public, most specialized interest groups tend to operate outside the electoral framework. It is useful
to consider, however, the circumstances that move a
given set of issues into the electoral arena, and
foreign policy provides two examples -- the Korean
War and the Vietnam War. At times when foreign policy issues are not salient to the majority of adults,
the foreign policy community operates much like the
scientific community, establishing regular channels
of expert communication from various groups to appropriate congressional committees and executive
agencies. The foreign policy elites and most of the
members of the attentive public for foreign policy
would assert that a high level of specialized knowledge is required to understand international diplomacy and that considerable effort must be devoted to
keeping informed about foreign affairs. The decision makers generally accept this proposition and
utilize established experts in the review and formation of foreign policy.
     In both the Korean and Vietnam conflicts, how-
ever, extended involvement by the United States,
including American troops and fatalities, began to
generate a division within the attentive public and,
ultimately, served to swell the size of the atten-

tive public for foreign policy. As the controversy grew, candidates began to make the war a principal issue, and numerous elections to the Congress were decided on foreign policy issues. Most political observers would agree that the Presidential elections of 1952 and 1968 were strongly influenced, if not directly decided, by the war issue. Two interesting trends emerged.

First, as public concern grew in each case, the complexity of the issues was narrowed. The issue was no longer framed as the appropriate American policy for Asia, or even Southeast Asia, but rather whether it was desirable to continue the national involvement in a particular armed conflict at the then current level. The simplification of the issue made it more susceptible to electoral determination.

Second, as the issues were simplified, the knowledge requirements for effective participation in the attentive public for foreign policy declined, encouraging greater participation. While much of the national media debate about the Vietnam War, for example, continued to be conducted by persons recognized as bringing a certain level of foreign policy expertise to the subject, the more common, and perhaps more effective, day-to-day debate among citizens was conducted by a wide range of involved and concerned nonexperts -- clergymen, student leaders, parents of soldiers, and business and labor leaders. In short, the minimum standards for both elites and attentives dropped as the size of the attentive public grew more massive and as the forum moved toward electoral policies.

It is unclear whether a similar pattern could occur with a scientific issue. The emerging protest movement over nuclear power will be an interesting test of this parallel between science policy and foreign policy. The leadership of the protest movement appears to seek a broader public involvement in the issue and to make nuclear power into an issue on which congressional and other candidates will have to take a public stand. The protest leaders expect to produce financial assistance and volunteers to those candidates opposing nuclear power. In contrast, the majority opinion within the scientific community would appear to favor nuclear power at the present time and to prefer to keep the issue out of the electoral process. It is a classic situation in which a group which has not been able to gain sufficient support within the narrower confines of its specialized community makes an appeal to the broader

public, sharing its preferred policy position. It
is too early to make an informed judgment about the
probable path of the nuclear power movement, but it
is illustrative of one type of force seeking greater
electoral involvement in scientific and technological issues.

Referenda: A Special Case

One-issue referenda represent an exception to the
general discussion above. For the most part, specialized interest communities have preferred to deal
with their problems outside the electoral arena and,
in the case of foreign policy, the alternative approach has been a general broadening of the issue
and the movement of the issue into regular electoral
contests for public office. Since foreign policy is
a national governmental function, referenda in state
or local elections could only be advisory, and the
return on the investment of political capital needed
would be minimal. Thus there were fewer referenda
on either the Korean or the Vietnam Wars.
     In contrast to foreign policy, issues involving
organized science appear to be much more susceptible
to referenda, as demonstrated in the cases of fluoridation and nuclear power. Since state and local
governments do have important legislative authorities over these matters, a successful referendum
would have the effect of bypassing the legislature
and the normal paths of influence exercised by specialized interest communities. Further, in those
states in which elections use a high level of media,
it may be possible to simplify a single issue like
nuclear power sufficiently to generate broader public interest and to influence a public decision to
act on the issue. The relatively large number of
referenda on nuclear power in recent years signals
the belief by the antinuclear movement that a solution is not possible -- at least in the short run --
within the scientific community and that it is necessary to expand the public's role in the issue.
     In many states, the number of signatures re-
quired to place an issue on a referendum ballot is
relatively low, and various groups have been able to
attain the necessary level of signatures for this
purpose. The process successfully circumvents the
influence paths of the specialized communities, and,
in the process, places the issue before the group

least interested and least informed in the topic --
the mass public. While the intent of groups following this strategy is to conduct a campaign to "edu-
cate" the public on that particular issue, many
groups have discovered that nonattentiveness is not
accidental, but reflects a deeply rooted interest
structure, and that it is most difficult to stimulate the level of issue interest and information
needed for a reasonable decision on most scientific
matters. Thus, important scientific questions may
be decided on the basis of whether more voters trust
John Wayne and Ronald Reagan than trust Paul Newman
and Jane Fonda.

In assessing the probable impact of the referendum strategy, it is useful to reflect on the attitudes of the nonattentive respondents discussed in
chapter eight. At the present time, the mass public
continues to hold the scientific and technological
communities in relatively high regard, although
there are some signs of wariness on selected issues.
In general, the mass public will approach most scientific issues with a positive outlook, and it will
be necessary for the challengers to persuade a
largely uninterested electorate that the issue is
important and that prevailing policy is wrong. To
date, the antinuclear movement has not found the
referendum to be a very successful strategy, but if
there are repeated incidents like Three Mile Island
and if a significant number of people become worried
about the possible health consequences of nuclear
leaks and accidents, nuclear power referenda may
become as numerous as were fluoridation referenda in
the 1950s and the early 1960s. It is a trend worth
watching.

THE ATTENTIVE PUBLIC FOR ORGANIZED SCIENCE

Given an analysis up to this point, what conclusions
are appropriate concerning the role of and the impact of the attentive public for organized science
within the American political system? Three major
conclusions emerge.

First, in comparison to the attentive publics
for foreign policy, economic policy, and civil
rights issues, the attentive public for organized
science is somewhat more populous than had been anticipated. Since science and technology issues are
relatively new entries on the political agenda in

comparison to the other three domains covered in the
1978 NPAS, it might have been expected that the
other areas would have been higher in interest and
awareness. Two explanations are possible. On the
one hand, the size could be discounted as a reflection of the higher than average exposure of students
to science courses in the environments of the high
school and college. This line of argument would
suggest that the level of attentiveness will decline
as the respondents leave formal schooling and move
into more normal work and family circumstances.

On the other hand, the relatively large size of
the attentive public could be seen as a reflection
of the growing prominence of scientific and technological issues in the political system, and as a
reflection of the differences in interests of the
newer generation. Dawson (1973) and others have
asserted that the traditional political and partisan
alignments are in "disarray," and it is possible to
argue that the relatively high level of young adult
interest in a traditionally nonpartisan issue cluster like organized science is but one more reflection of a basic rearrangement of the American
political agenda. Unfortunately, the data from the
1978 NPAS do not provide a basis for choosing between the explanations.

Second, the data indicate that when attentiveness to organized science is clustered with other
domains, it is most often clustered with economic
and civil rights concerns, and rarely with foreign
policy concerns. These clusters imply that many of
the respondents see science and technology as related to important domestic problems, or to the solution of these problems. This clustering may reflect
shared concerns about environmental conservation,
energy for the future, nuclear power, or other domestic issues. A concern for human rights would appear
to be related to a strong concern about the welfare
of individuals, recalling the factor analysis of
headline interests in chapter five.

Traditionally, the scientific community has attempted to separate scientific issues from other domestic and foreign problems, and by and large the
scientific community has been reluctant to promise
major scientific contributions to the persistent social and economic ills of modern industrial societies. The coupling of attentiveness to organized
science with concerns about domestic and humanitar-
ian issues may indicate a disposition on the part of
the attentive public that is, at least in part, in

conflict with the traditional views of the leadership of the scientific community.

Third, and perhaps most basic, this analysis of
the attentive public for organized science poses a
dilemma for the leadership of the scientific community -- the elites in the system. For the most
part, this group is strongly committed to a broader
"public understanding of science," but this public
understanding is often presumed to be one of passive
appreciation. It is the nature of attentive publics
to be somewhat more critical and not very passive.
After all, why would a person make the investment of
time and effort to become informed just to admire
from afar? Since many of the persons in the attentive public for organized science will not be
practicing scientists or engineers, but rather
interested and concerned citizens, the larger the
attentive public becomes, the greater will be the
"external" intervention in the policies of the scientific community. And, it will be recalled, the
prevention of external intervention has been one of
the basic objectives of the scientific community for
the last three decades -- thus, the dilemma.

Despite any reservations that might be held by
the elites in the scientific community, the data
analyzed above strongly suggest that the size of the
attentive public for organized science in the newest
political generation will be larger than the attentive public in previous age cohorts, and there are
substantial reasons for expecting the trend to continue. The impact of scientific and technological
activities on American society will undoubtedly increase. The current search for alternative energy
sources is but one of numerous examples. More
science-based controversies will come before the
Congress and state legislatures. The media have
demonstrated a growing commitment to covering science news, and numerous specialized information
channels designed for attentives are appearing. An
increasing proportion of the American people have a
college education, if not advanced degrees, and
these seem to be prerequisites to attentiveness.
Given these conditions, it is reasonable to expect
substantial growth in the attentive public for
organized science in coming decades, opening the way
for a larger policy-influencing role for interested
and concerned citizens.

Appendix A:
The 1978 NPAS
Instruments

The data for the 1978 National Public Affairs Study
were collected using a high school and a college
version of a basic instrument. The design of
the instrument was described in Chapter Four. This
appendix contains the high school version of the
instrument. The college version was modified to
reflect disciplinary descriptions common to college
level coursework (i.e., humanities instead of
English), to delete certain choices not applicable
to college students (i.e., less than high school
graduation as the highest level of education expected), and to delete teacher from the information
confidence questions since college professor was
already included.

Occupational codes used to complete question 6, School and Career Plans, and questions 6a and 6b, You and Your Family.

10 Accountant or actuary

11 Actor or entertainer

12 Architect or city planner

13 Artist or designer

14 Business executive or manager

15 Business owner or proprietor

16 Business salesman or buyer

17 Clery (minister, priest, rabbi)

18 Clinical psychologist

19 College teacher or administrator

20 Computer programmer or analyst

21 Conservationist or forester

22 Dentist

23 Dietitian or home economist

24 Elective office holder

25 Engineer

26 Farm or ranch owner or manager

27 Farm or ranch worker

28 Foreign service officer

29 Homemaker (full-time)

30 Interior decorator

31 Lab technician or hygienist

32 Laborer (unskilled)

33 Law enforcement officer

34 Lawyer, attorney, or judge

35 Librarian

36 Military service

37 Musician (performer, composer)

38 Nurse

39 Optometrist

40 Pharmacist

41 Physician

42 Public administrator

43 Radio/television production

44 Retail clerk

45 School counselor

46 Scientific researcher

47 Secretary

48 Semi-skilled worker

49 Skilled trades (carpenter, mechanic, etc.)

50 Social, welfare, or recreation worker

51 Student (full-time)

52 Teacher or administrator (elementary)

53 Teacher or administrator (secondary)

54 Transportation worker

55 Veterinarian

56 Writer or journalist

57 Unemployed

58 Retired

99 Does not apply

1978 NATIONAL PUBLIC AFFAIRS STUDY

1. In your opinion, what are the three most important issues or problems facing the United States today?

   a. _____

   b. _____

   c. _____

2. Would you say that you are **very** interested, **somewhat** interested, or **not very** interested in most news and current events issues? (circle one)

   Very interested 1
   Somewhat interested 2
   Not very interested 3

3. How often do you read a newspaper? (circle one)

   Every day 1
   5-6 times a week 2
   3-4 times a week 3
   1-2 times a week 4
   Not at all 5

4. When you read a newspaper, which sections do you usually read? (circle one answer for each line)

	Usually Read	**Usually Don't**
a. Sports	1	2
b. Fashion	1	2
c. Local news	1	2
d. State news	1	2
e. National news	1	2
f. Editorials	1	2

5. How often do you watch a television newscast? (circle one)

   Every day 1
   5-6 times a week 2
   3-4 times a week 3
   1-2 times a week 4
   Not at all 5

6. Please list the magazines that you read regularly, that is, almost every issue. Include all magazines regardless of their news content.

   a. _____ b. _____ c. _____

   d. _____ e. _____ f. _____

7. Please list the magazines that you read occasionally, that is, some of the issues but not most issues. Include all magazines regardless of their news content.

   a. _____ b. _____ c. _____

8. Which of the following is your most important source of information about current news events? (circle one)

   Radio 1
   Television 2
   Newspapers 3
   Magazines 4
   Other (please specify: _____) 5

9. We are interested in the types of news stories you find most interesting and would be likely to read. Below are listed several sets of headlines similar to those you might find in a newspaper or magazine. For each headline, please indicate whether you would **definitely** (D) read it, **probably** (P) read it, **probably not** (PN) read it, or **definitely not** (DN) read it by circling the appropriate letter(s). Consider the constraints you have on your time and be as realistic as possible. Please read all headlines once before you start making responses.

Definitely not read ⟶				**Definitely not read** ⟶				
Probably not read ⟶				**Probably not read** ⟶				
Probably read ⟶				**Probably read** ⟶				
Definitely read ⟶				**Definitely read** ⟶				

Headline	D	P	PN	DN	Headline	D	P	PN	DN
New Italian Elections May Change NATO Alliance	D	P	PN	DN	President To Ask Congress To Create New Jobs	D	P	PN	DN
Baseball Writers Set Favorites For New Season	D	P	PN	DN	Nutritionist Questions Value Of Most Breakfast Cereals	D	P	PN	DN
Senator Asserts New Policy Needed To Fight Inflation	D	P	PN	DN	Medical Team Reports New Birth Control Device	D	P	PN	DN
Scientist Explains Research In Human Cell Modification	D	P	PN	DN	Supreme Court Rules In Freedom Of Speech Case	D	P	PN	DN
Soviets Launch New Unmanned Space Station	D	P	PN	DN	U.S. Scientists Assert Need For Research Funding	D	P	PN	DN
State Court Rules On Sex Discrimination	D	P	PN	DN	New Arab-Israeli Talks Hint Middle East Peace	D	P	PN	DN
Defense Secretary Reveals New Long-Range Missile	D	P	PN	DN	Cossell Reviews Strength Of NFL For Next Season	D	P	PN	DN
Oil Company Executive Assesses Energy Crisis	D	P	PN	DN	Space Scientist Asserts Need For Manned Flight	D	P	PN	DN
Scientific Group Reports On Great Lakes Pollution	D	P	PN	DN	Experts Assess Need For Basic Welfare Reform	D	P	PN	DN
Third-World Nations Seek New Trade Treaties	D	P	PN	DN	Professor Reports National Study Of UFO Sightings	D	P	PN	DN
Sociologist Describes Impact Of Unemployment On Family	D	P	PN	DN	Geologists Release New Report On Oil Reserves	D	P	PN	DN
Suit Filed To Test Affirmative-Action Rules	D	P	PN	DN	European Common Market Sets Foreign Trade Policy	D	P	PN	DN
President Releases Report On Basic Science Research	D	P	PN	DN	University Scientists Close To Cancer Therapy Drug	D	P	PN	DN
Paris And London Editors Preview Summer Fashions	D	P	PN	DN	UN Secretary General Reviews Conflict In Southern Africa	D	P	PN	DN
Congress Moves To Curb Chemical Pollution	D	P	PN	DN	Air Force To Test New Weapons System	D	P	PN	DN
Senate Witnesses Debate Future Of Solar Energy	D	P	PN	DN	Judge To Rule On Police Search Procedures	D	P	PN	DN

FOREIGN AFFAIRS

1. Generally speaking, would you say that you are **very** interested, **somewhat** interested, or **not very** interested in foreign policy issues? (circle one)

   Very interested 1
   Somewhat interested 2
   Not very interested 3

2. Below are listed several statements involving foreign policy. Please indicate if you **agree** (A) or **disagree** (D) by circling the appropriate letter. If you **strongly agree** (SA) or **strongly disagree** (SD) with a statement, circle the appropriate letters. If you are **uncertain** how you feel about this statement, please circle the letter U.

Uncertain ────────────────────────
Strongly disagree ──────────────────
Disagree ──────────────
Agree ──────────
Strongly agree ──────────

a. By the year 2000, the U.S. should turn over control of the Panama Canal
   to the government of Panama.   SA   A   D   SD   U

b. The United States and Cuba should establish formal diplomatic relations.   SA   A   D   SD   U

c. Most citizens are not well enough informed to make useful input to
   foreign policy decisions.   SA   A   D   SD   U

d. The United States should place less emphasis on military force and
   more emphasis on diplomacy.   SA   A   D   SD   U

e. The United States should refuse to cooperate economically or politically
   with Communist countries.   SA   A   D   SD   U

f. Decisions about foreign policy issues should be left up to diplomats
   and other foreign policy experts.   SA   A   D   SD   U

g. If there is another Arab oil boycott, the United States should send
   in troops and seize the wells.   SA   A   D   SD   U

h. The United States should continue to participate actively in the
   United Nations.   SA   A   D   SD   U

i. The United States should not cooperate with or recognize dictatorships
   in any nation.   SA   A   D   SD   U

j. The interested and informed citizen can often have some influence on
   foreign policy decisions if he is willing to make the effort.   SA   A   D   SD   U

k. The United States should pay some compensation to Vietnam
   for damages caused during the recent war.   SA   A   D   SD   U

l. The United States should actively encourage other nations to adopt
   democratic forms of government.   SA   A   D   SD   U

m. The United States should provide more foreign economic aid to
   underdeveloped countries.   SA   A   D   SD   U

n. Foreign policy decisions should be based primarily on moral and
   human rights considerations.   SA   A   D   SD   U

3. Please provide a brief answer to each of the questions:

   a. Who is the Prime Minister of Great Britain? _____

   b. Who is Secretary General of the United Nations now? _____

   c. What is NATO? _____

   d. What is meant by "third-world nations"? _____

   e. What is OPEC? _____

4. So far this school year, how many times have you had a discussion of a current foreign policy issue in your classes? (circle one)

   None 1
   One or two times 2
   Three or four times 3
   Five or more times 4

5. So far this school year, how many times have you had a discussion of a current foreign policy issue with another student outside of class? (circle one)

   None 1
   One or two times 2
   Three or four times 3
   Five or more times 4

6. Thinking back to this school year, how many times have you had a discussion of a current foreign policy issue with members of your family? (circle one)

   None 1
   One or two times 2
   Three or four times 3
   Five or more times 4

7. How much would you trust each of the following people or sources to give you accurate and truthful information about **foreign policy** issues? (check one response for each line)

	A lot	Some	Not much	Not sure
a. The TV evening news	☐	☐	☐	☐
b. President Carter	☐	☐	☐	☐
c. A university professor	☐	☐	☐	☐
d. The U.S. State Department	☐	☐	☐	☐
e. A radio news broadcast	☐	☐	☐	☐
f. A student in your school	☐	☐	☐	☐
g. **Newsweek** or **Time**	☐	☐	☐	☐
h. Your father	☐	☐	☐	☐
i. Your mother	☐	☐	☐	☐
j. A U.S. military leader	☐	☐	☐	☐
k. The United Nations	☐	☐	☐	☐
l. The Senate Foreign Relations Committee	☐	☐	☐	☐
m. A teacher in your school	☐	☐	☐	☐

SCIENCE AND TECHNOLOGY

1. Generally speaking, would you say that you are **very** interested, **somewhat** interested, or **not very** interested in issues involving science and technology? (circle one)

Very interested 1
Somewhat interested 2
Not very interested 3

2. Below are listed several statements involving science and technology. Please indicate if you **agree** (A), or **disagree** (D) by circling the appropriate letter. If you **strongly agree** (SA) or **strongly disagree** (SD) with a statement, circle the appropriate letters. If you are **uncertain** how you feel about a statement, please circle the letter U.

Uncertain
Strongly disagree
Disagree
Agree
Strongly agree

a. Scientific invention is largely responsible for our standard of living in the United States. SA A D SD U

b. There are still major unanswered questions about the safety of nuclear power plants. SA A D SD U

c. Most citizens are not well enough informed to make useful input to policy decisions concerning science and technology. SA A D SD U

d. It is important that the U.S. continue to develop new weapons at least as fast as the Russians. SA A D SD U

e. We can depend on science and technology for a long-term solution to the energy problem. SA A D SD U

f. Decisions about science policy issues should be left up to scientists and other science policy experts. SA A D SD U

g. A privately-owned chemical company should be allowed to manufacture and sell anything they want without governmental interference. SA A D SD U

h. Solar energy is the best single long-term solution to our energy problem. SA A D SD U

i. One of the bad effects of science is that it breaks down people's ideas of right and wrong. SA A D SD U

j. The interested and informed citizen can often have some influence on science policy decisions if he is willing to make the effort. SA A D SD U

k. The risk involved in generating nuclear power is relatively minor and should not block the construction of new nuclear power plants. SA A D SD U

l. The growth of science means that a few people could control our lives. SA A D SD U

m. Overall, science and technology have caused more good than harm. SA A D SD U

n. One trouble with science is that it makes our lives change too fast. SA A D SD U

o. The Federal Government should spend more money on science research. SA A D SD U

p. One of our big troubles is that we depend too much on science and not enough on faith. SA A D SD U

q. Science is making our lives healthier, easier, and more comfortable. SA A D SD U

r. When all is said and done, the space program still hasn't produced much of value for this country. .SA A D SD U

s. The U.S. has lost its lead in science research to the Soviet Union and other nations. SA A D SD U

t. Science policy decisions should be based primarily on moral and human rights considerations. SA A D SD U

3. Please provide a brief answer to each of the questions:

    a. What is a molecule? _____

        _____

    b. What is an organic chemical? _____

        _____

    c. What is an amoeba? _____

        _____

    d. What is DNA? _____

        _____

    e. Some people want to build more nuclear power plants. What arguments have you heard in favor of this position?

        _____

        _____

        _____

    f. Some people want to prohibit more nuclear power plants. What arguments have you heard in favor of this position?

        _____

        _____

4. So far this school year, how many times have you had a discussion of a current science related issue in your classes? (circle one)

                    None 1
                    One or two times 2
                    Three or four times 3
                    Five or more times 4

5. So far this school year, how many times have you had a discussion of a current science related issue with another student outside of class? (circle one)

                    None 1
                    One or two times 2
                    Three or four times 3
                    Five or more times 4

6. Thinking back to this school year, how many times have you had a discussion of a current science related issue with members of your family? (circle one)

                    None 1
                    One or two times 2
                    Three or four times 3
                    Five or more times 4

7. How much would you trust each of the following people or sources to give you accurate and truthful information about **science policy** issues? (check one response for each line)

	A lot	Some	Not much	Not sure
a. The TV evening news	□	□	□	□
b. President Carter	□	□	□	□
c. A university professor	□	□	□	□
d. The Environmental Protection Agency	□	□	□	□
e. A radio news broadcast	□	□	□	□
f. A student in your school	□	□	□	□
g. **Newsweek** or **Time**	□	□	□	□
h. Your father	□	□	□	□
i. Your mother	□	□	□	□
j. The president of a chemical company	□	□	□	□
k. The United Nations	□	□	□	□
l. A Congressional Committee on Science and Technology	□	□	□	□
m. A teacher in your school	□	□	□	□

8. Listed below are several science and technology areas for which federal taxes are sometimes used. Please select three activities that you would most like to have supported with tax funds and the three activities that you would least like to have supported with tax funds. (please check **three** boxes in the **most like** column and **three** boxes in the **least like** column)

	Most like	Least like
Improving the safety of automobiles	□	□
Reducing crime	□	□
Discovering new basic knowledge about man and nature	□	□
Developing or improving weapons for national defense	□	□
Improving education	□	□
Reducing and controlling pollution	□	□
Finding better birth control methods	□	□
Space exploration	□	□
Finding new methods for preventing and treating drug addiction	□	□
Improving heath care	□	□
Developing faster and safer public transportation	□	□
Developing new energy sources	□	□
Weather control and prediction	□	□

ECONOMIC ISSUES

1. Generally speaking, would you say that you are **very** interested, **somewhat** interested, or **not very** interested in economic issues like unemployment and inflation?

    Very interested 1
    Somewhat interested 2
    Not very interested 3

2. Below are listed several statements involving economic issues. Please indicate if you **agree** (A) or **disagree** (D) by circling the appropriate letter. If you **strongly agree** (SA) or **strongly disagree** (SD) with a statement, circle the appropriate letters. If you are **uncertain** how you feel about a statement, please circle the letter U.

                                                    Uncertain ────────────────────────────
                                            Strongly disagree ──────────────────
                                                    Disagree ──────────────
                                                       Agree ──────────
                                             Strongly agree ──────

    a. Most of the people on unemployment could find jobs if they really wanted to.   SA   A   D   SD   U

    b. To stop inflation, the government should place controls on wages and prices.   SA   A   D   SD   U

    c. Most citizens are not well enough informed to make useful input to economic policy decisions.   SA   A   D   SD   U

    d. People with higher incomes should pay a higher **rate** of taxes than people with lower incomes.   SA   A   D   SD   U

    e. If we have to choose, it is more important to reduce inflation than to reduce unemployment.   SA   A   D   SD   U

    f. Decisions about economic policy issues should be left up to economists and other economic policy experts.   SA   A   D   SD   U

    g. The recent increases in prices are caused mainly by the desire of corporations for higher profits.   SA   A   D   SD   U

    h. If the imported goods from another country are hurting an American industry, the government should prohibit or limit the importation of that group of foreign products.   SA   S   D   SD   U

    i. The federal income tax rate discourages individual hard work today.   SA   A   D   SD   U

    j. The interested and informed citizen can often have some influence on economic policy decisions if he is willing to make the effort.   SA   A   D   SD   U

    k. Unions have done more good than harm for the American economy.   SA   A   D   SD   U

    l. The federal government should be required to balance its budget **every** year.   SA   A   D   SD   U

    m. The increased price of oil and energy is the principal cause of inflation today.   SA   A   D   SD   U

    n. Economic policy decisions should be based primarily on moral and human rights considerations.   SA   A   D   SD   U

3. Please provide a brief answer to each of the questions.

    a. What is meant by the GNP? _____

        _____

    b. What is inflation? _____

        _____

c. What is the present unemployment rate in the United States? _____ %

d. What is meant by "the balance of payments?" _____

_____

e. Who is the Secretary of the U.S. Department of Health, Education, and Welfare?

_____

4. So far this school year, how many times have you had a discussion of a current economic policy issue in your classes? (circle one)

       None 1
       One or two times 2
       Three or four times 3
       Five or more times 4

5. So far this school year, how many times have you had a discussion of a current economic policy issue with another student outside of class? (circle one)

       None 1
       One or two times 2
       Three or four times 3
       Five or more times 4

6. Thinking back to this school year, how many times have you had a discussion of a current economic policy issue with members of your family? (circle one)

       None 1
       One or two times 2
       Three or four times 3
       Five or more times 4

7. How much would you trust each of the following people or sources to give you accurate and truthful information about **economic policy** issues? (check one response for each line)

		A lot	Some	Not much	Not sure
a.	The TV evening news	☐	☐	☐	☐
b.	President Carter	☐	☐	☐	☐
c.	A university professor	☐	☐	☐	☐
d.	The U.S. Commerce Department	☐	☐	☐	☐
e.	A radio news broadcast	☐	☐	☐	☐
f.	A student in your school	☐	☐	☐	☐
g.	**Newsweek** or **Time**	☐	☐	☐	☐
h.	Your father	☐	☐	☐	☐
i.	Your mother	☐	☐	☐	☐
j.	A business leader	☐	☐	☐	☐
k.	The Senate Commerce Committee	☐	☐	☐	☐
l.	The United Nations	☐	☐	☐	☐
m.	A teacher in your school	☐	☐	☐	☐

CIVIL RIGHTS

1. Generally speaking, would you say that you are **very** interested, **somewhat** interested, or **not very** interested in civil rights issues? (circle one)

    Very interested 1
    Somewhat interested 2
    Not very interested 3

2. Below are listed several statements about civil rights issues. Please indicate if you **agree** (A) or **disagree** (D) by circling the appropriate letter. If you **strongly agree** (SA) or **strongly disagree** (SD) with a statement, circle the appropriate letters. If you are **uncertain** how you feel about a statement, please circle the letter U.

Uncertain
Strongly disagree
Disagree
Agree
Strongly agree

		SA	A	D	SD	U
a.	The Constitution should be amended to provide for equal rights for women.	SA	A	D	SD	U
b.	The courts are too easy on criminal defendants.	SA	A	D	SD	U
c.	Most citizens are not well enough informed to make useful input to civil rights decisions.	SA	A	D	SD	U
d.	The government should be allowed to stop the publication of information that it believes would be harmful to the national interest.	SA	A	D	SD	U
e.	The idea of quota systems for minorities or women is a violation of the Constitution.	SA	A	D	SD	U
f.	Decisions about civil rights issues should be left up to the judges and the courts.	SA	A	D	SD	U
g.	The Constitution should be amended to allow for prayers in the public schools.	SA	A	D	SD	U
h.	The interested and informed citizen can often have some influence on civil rights decisions if he is willing to make the effort.	SA	A	D	SD	U
i.	The present Supreme Court is too liberal on civil rights issues.	SA	A	D	SD	U
j.	When people seriously disagree with a law, they should follow their consciences instead of the law.	SA	A	D	SD	U
k.	Civil rights decisions should be based primarily on the United States Constitution.	SA	A	D	SD	U
l.	The police should be able to tap the phones of persons they suspect of committing crimes.	SA	A	D	SD	U
m.	The right to freedom of speech is absolute, regardless of what a person is saying.	SA	A	D	SD	U
n.	It is a mistake for the courts to require the busing of students for purposes of racial balance in the schools.	SA	A	D	SD	U
o.	Civil rights decisions should be based primarily on moral and human rights considerations.	SA	A	D	SD	U

3. Please provide a brief answer to each of the questions.

    a. What is the ACLU? _____

    _____

b. When there is a disagreement about the meaning of the U.S. Constitution, how is the dispute resolved?

_____

c. What is the Bill of Rights? _____

d. What is the Equal Rights Amendment? _____

e. What is meant by "Fifth Amendment" protection? _____

_____

4. So far this school year, how many times have you had a discussion of a civil rights issue in your classes? (circle one)

    None 1
    One or two times 2
    Three or four times 3
    Five or more times 4

5. So far this school year, how many times have you had a discussion of a civil rights issue with another student outside of class? (circle one)

    None 1
    One or two times 2
    Three or four times 3
    Five or more times 4

6. Thinking back to this school year, how many times have you had a discussion of a civil rights issue with members of your family? (circle one)

    None 1
    One or two times 2
    Three or four times 3
    Five or more times 4

7. How much would you trust each of the following people or sources to give you accurate and truthful information about **civil rights** issues? (check one response for each line)

		A lot	Some	Not much	Not sure
a.	The TV evening news	☐	☐	☐	☐
b.	President Carter	☐	☐	☐	☐
c.	A university professor	☐	☐	☐	☐
d.	The U.S. Justice Department	☐	☐	☐	☐
e.	A radio news broadcast	☐	☐	☐	☐
f.	A student from your school	☐	☐	☐	☐
g.	**Newsweek** or **Time**	☐	☐	☐	☐
h.	Your father	☐	☐	☐	☐
i.	Your mother	☐	☐	☐	☐
j.	A federal judge	☐	☐	☐	☐
k.	The United Nations	☐	☐	☐	☐
l.	A Senate Committee	☐	☐	☐	☐
m.	A teacher in your school	☐	☐	☐	☐

PEOPLE AND POLITICS

Much of what happens in public affairs involves people in government. In this section, we would like to know your opinions about the way government officials do their jobs.

1. Would you say that most of the time people try to be helpful, or that they are mostly looking out for themselves? (circle one)

    Try to be helpful most of the time 1
    Just looking out for themselves 2
    I have no opinion 3

2. Do you think that most people try to take advantage of you if they get the chance, or do most people try to be fair? (circle one)

    Most try to take advantage 1
    Most try to be fair 2
    I have no opinion 3

3. Generally speaking, would you say that most people can be trusted, or that you can't be too careful in dealing with people? (circle one)

    Most people can be trusted 1
    You can't be too careful 2
    I have no opinion 3

4. Thinking of the people who run the government, how many would you say are crooked? (circle one)

    Quite a few 1
    Not very many 2
    I have no opinion 3

5. Do you think that the people who run the government waste a lot of tax money, some tax money, or very little tax money? (circle one)

    A lot 1
    Some 2
    Very little 3
    I have no opinion 4

6. How much of the time do you think you can trust the government to do what is right? (circle one)

    Almost always 1
    Most of the time 2
    Only some of the time 3
    Almost never 4
    I have no opinion 5

7. How competent are the people who are running the government? (circle one)

    Most are very competent 1
    Many are somewhat incompetent 2
    Most don't know what they are doing 3
    I have no opinion 4

8. How much attention do you feel the government pays to what people think when it decides what to do? (circle one)

    A lot 1
    Some 2
    Not much 3
    I have no opinion 4

Individuals' interests in public affairs are often related to their opinions on more general issues. We would like to know your opinions regarding the issues listed below. Please circle one answer for **each** statement.

Uncertain ────────────────────┐
Strongly disagree ──────────┐ │
Disagree ──────────┐ │ │
Agree ──────┐ │ │ │
Strongly agree ─┐ │ │ │ │

9. There will always be wars, no matter how hard people try to prevent them. SA A D SD U

10. Fundamentally, the world we live in is a pretty lonesome place. SA A D SD U

11. In the long run, the best way to live is to pick friends and associates whose tastes and beliefs are the same as your own. SA A D SD U

12. Is it not always wise to plan too far ahead because many things turn out to be a matter of good or bad fortune anyhow. SA A D SD U

13. Most people just don't give a damn for others. SA A D SD U

14. It is only when a person devotes himself to an ideal or cause that life becomes meaningful. SA A D SD U

15. People can be divided into two classes: the weak and the strong. SA A D SD U

16. By taking an active part in political and social affairs, the people can control world events. SA A D SD U

17. Of all the philosophies in the world, there is probably only one which is correct. SA A D SD U

18. As far as world affairs are concerned, most of us are the victims of forces we can neither understand, nor control. SA A D SD U

19. It is only natural that a person should have a much better acquaintance with ideas he believes in than ideas he opposes. SA A D SD U

20. Strong discipline builds strong character. SA A D SD U

21. In this complicated world of ours, the only way we can know what is going on is to rely on leaders and experts who can be trusted. SA A D SD U

22. Without the right breaks one cannot be an effective leader. SA A D SD U

23. A person who has not believed in a great cause has not really lived. SA A D SD U

24. When it comes to differences in opinion, we must be careful not to compromise with those who believe differently than we do. SA A D SD U

25. Getting a good job depends mainly on being in the right place at the right time. SA A D SD U

26. Generally speaking, the party preferences of me and my parents are: (circle one in each column)

	Me	My Mother	My Father
A strong republican	1	1	1
A republican	2	2	2
An independent republican	3	3	3
An independent	4	4	4
An independent democrat	5	5	5
A democrat	6	6	6
A strong democrat	7	7	7
A member of another political party	8	8	8
None of the above (including deceased)	9	9	9

PERSONAL AND RELIGIOUS VALUES

In the preceding sections you have told us a great deal about your interests in current affairs. To help us make meaningful comparisons among people with different backgrounds and aspirations, the last three sections ask for information about you — your religious beliefs, your educational and occupational aspirations, and some questions about your family. Please remember all replies will be held in strict confidence.

1. Which of the following statements most closely represent your beliefs? Please answer **each** question indicating how strongly you agree or disagree with the statement.

Uncertain ─────────────────
Strongly disagree ──────────
Disagree ──────────
Agree ─────────
Strongly agree ────

		SA	A	D	SD	U
a.	I believe there is life after death	SA	A	D	SD	U
b.	I believe there is a God who can and has personally intervened in the lives of men and women	SA	A	D	SD	U
c.	I believe that the Bible is literally true and we should believe everything it says.	SA	A	D	SD	U
d.	I feel it is important for my children to believe in God	SA	A	D	SD	U
e.	Prayer plays an important role in my life	SA	A	D	SD	U
f.	There may be a God and there may not be	SA	A	D	SD	U
g.	Faith in the supernatural is a harmful self-delusion	SA	A	D	SD	U

2. What is your religious affiliation? (circle one)

Protestant (please specify: _____) 1
Roman Catholic 2
Jewish 3
Other (please specify: _____) 4
None 5

3. How often have you attended religious services so far this school year? (circle one)

More than once a week 1
Every week 2
Two or three times each month 3
Four to 10 times during the year 4
Three times or fewer during the year 5
Never 6

SCHOOL AND CAREER PLANS

1. What year are you in school? (circle one)

Freshman 1
Sophomore 2
Junior 3
Senior 4

2. What is your overall high school average? (circle one)

D- to D (0.5-0.9) 1
D to C- (1.0-1.4) 2
C- to C (1.5-1.9) 3
C to B- (2.0-2.4) 4
B- to B (2.5-2.9) 5
B to B+ (3.0 to 3.4) 6
A- to A (3.5-4.0) 7

3. As best you know, how does your grade average rank among students in your school? (circle one)

    Top 5% 1
    Next 5% 2
    Upper quarter, but not top 10% 3
    Upper half 4
    Lower half 5

4. Please indicate the number of semesters (including the present semester) that you have completed in each of the following subjects or groups of subjects.

English .... ☐

Mathematics .... ☐

Biology (including botany and zoology) .... ☐

Chemistry .... ☐

Physics (including physical science) .... ☐

Other natural sciences
(general science, earth science, etc.) .... ☐

Social sciences (civics, history, economics, etc.) . ☐

Foreign language .... ☐

Business-commercial .... ☐

Vocational-occupational .... ☐

5. What is the highest level of education you expect to complete? (circle one)

    High school 1
    Vocational or technical program (less than 2 years) 2
    Two-year college degree 3
    Bachelor's degree 4
    One or 2-year graduate degree (MA, MS, etc.) 5
    Professional degree beyond 2-years (PhD, MD, etc.) 6
    Other (please specify: _____) 7

6. What do you expect your occupation (vocation) to be? Use the list of occupations inserted in this booklet, and write the **number** of the occupation in the boxes below. If you have absolutely no idea what occupation you expect to be in, write the number "99" in the two boxes.

    ☐☐

7. How are you of this occupational choice? (circle one)

    Very sure 1
    Fairly sure 2
    Not at all sure 3

8. Looking ahead into the future, how important would you consider each of the following? (Please circle one response for **each** line)

	Essential	Very important	Somewhat important	Not important	Uncertain
a. Becoming accomplished in my career field	E	V	S	N	U
b. Influencing political decisions and structures	E	V	S	N	U
c. Developing a meaningful philosophy of life	E	V	S	N	U
d. Participating in community affairs	E	V	S	N	U
e. Raising a family	E	V	S	N	U
f. Becoming well off financially	E	V	S	N	U
g. Becoming an authority in my field	E	V	S	N	U
h. Having administrative responsibility over others' work	E	V	S	N	U
i. Influencing social values	E	V	S	N	U
j. Obtaining recognition from colleagues for contributions to my special field	E	V	S	N	U
k. Keeping up to date with political affairs	E	V	S	N	U

YOU AND YOUR FAMILY

In this final section, we want to learn something more about you as a person. Please remember that all replies will be held in strict confidence.

1. What is your sex? (circle one)   Female 1
                                        Male 2

2. Please list the age of any brothers you may have from oldest to youngest.
    oldest □□   □□   □□   □□   □□   □□   □□   □□

3. Please list the age of any sisters you may have from oldest to youngest.
    oldest □□   □□   □□   □□   □□   □□   □□   □□

4. What is your present age? □□

5. What is the highest level of education obtained by your parents? (circle the highest obtained in **each** column)

	Father	Mother
Grammar school or less	0	0
Some high school	1	1
High school graduate	2	2
Post-high school degree		
(technical, business, or community college)	3	3
Some college	4	4
College graduate	5	5
One or 2-year graduate degree (MA, MBA, etc.)	6	6
Higher graduate or professional degree (PhD, MD, etc.)	7	7
Other (please specify: _____)	8	8
I don't know	9	9

6. On the sheet that is inserted in your booklet you will find a list of occupations numbered from 10 to 99 organized into different fields of work. Read over this list carefully to find your parents occupations, then write in the **number** corresponding to these occupations in the boxes below. If their occupations do not appear on the list, write the name of the occupation next to the appropriate boxes.

    a. **What is (or was) your father's occupation?** □□ _____

    b. **What is (or was) your mother's occupation?** □□ _____

7. The statements listed below are often used to describe people. Please indicate if each statement sounds **very much** like you (V), **somewhat like you** (S), or **not at all** like you (N). Please circle one response for each statement.

    Not at all like me ────────┐
    Somewhat like me ──────┐   │
    Very much like me──┐   │   │

		V	S	N
a.	When I express an opinion, others usually follow my suggestions	V	S	N
b.	There are lots of things about me I would change if I could	V	S	N
c.	I don't really enjoy school, but I feel that I must go in order to have the things I need and want later in life	V	S	N
d.	I am not much interested in the TV programs, movies, or magazines that people my age seem to like	V	S	N
e.	I find it hard to talk in front of a group	V	S	N
f.	Sometimes I feel I don't have enough control over the direction my life is taking ...	V	S	N
g.	If given a chance, I would do something of great benefit to the world	V	S	N
h.	I am popular with people my own age	V	S	N
i.	I usually give in very easily when people disagree with me	V	S	N

We appreciate the time and effort you have put into answering this questionnaire. Thank you.

Appendix B:
Log-linear Analysis—
Logic and Procedures

Log-linear models are used extensively in the present study to describe differences in attentiveness
to science among subsamples of the NPAS survey and
to make inferences about the characteristics of the
student populations from which these subsamples were
drawn. This appendix discusses the logic and procedures used in constructing these models.

ACCOUNTING FOR VARIATION AMONG QUALITATIVE VARIABLES

When a sample is partitioned across multiple dimensions, the problem for the analyst is to determine
how much variation is associated with differences
among each of the qualitatively distinct subsamples.
There are as many potential sources of variation as
there are degrees of freedom in a given contingency
table. Variation across each marginal distribution
and association among marginals contributes to the
total variation among subsamples.
    Table B.1 is adapted from Fig. 9.2 in the text
to illustrate the use of log-linear models. Table
B.1 partitions the total sample of 3738 students
responding to questions about attentiveness to science, occupational aspirations, and educational status into 18 subsamples. The sources of potential
variation are given with their respective degrees of
freedom in Table B.2.
    To attribute variation among the subsamples to
different sources of variation, the analyst constructs a series, or "hierarchy", of models incorpo-

Table B.1 Attentiveness to Organized Science by
Occupational Aspirations and Educational Cohort

Group	Occupational Aspirations	Attentiveness to Science	
		Nonattentive	Attentive
High school- Non-college- Bound	Undecided Medium/Low High	339 771 53	6 46 3
High School- College-Bound	Undecided Medium/Low High	180 551 181	23 120 42
College	Undecided Medium/Low High	120 692 193	59 241 118
Total		3080	658

rating different sets of marginals (see Table B.3).
Using each model as a predictive equation, "expected
values" are predicted for each subsample and compared to the observed frequencies. Discrepancies between expected and observed frequencies indicate
that the model on which the expected values are
based is inadequate to account for all the variation
in the table. In this case, additional marginals
are incorporated until a parsimonious model is constructed that accounts for the variation in the
table within acceptable levels of sampling error.
    The most parsimonious model that can be con-
structed uses only the sample size (N) to predict
the frequencies in each subsample. This "null model" states that all of the subpopulations from which
the respondents were drawn contain equal numbers of
people, or, in other words, there are no differences
among categories of the univariate marginals nor any
associations among these marginals in the population.
    The expected value under the assumption of no
differences among the subsamples is obtained by dividing the sample size (N) by the number of cells in
the table (k). The result (3738/18 = 207.67) indicates that, if there were no differences in the

representation of these 18 subpopulations in the
sample, one would expect 207.67 respondents to be
found in each cell. To the extent that the observed
values differ from the expected, one can conclude
that certain combinations of characteristics are
more likely to appear together than are others.
     The likelihood ratio chi-square (L.R. χ^2) is a
measure of the extent of discrepancy between the observed frequencies and those that would be expected
if the model fit the data (Feinberg 1977):

$$\text{L.R. } \chi^2 = 2 \sum_{i}^{k} \text{observed ln (observed/expected)} \quad (1)$$

The large L.R. χ^2 associated with the null model in
Table B.3 indicates that the value 207.67 provides a
poor fit to the observed subsample frequencies in
Table B.1. Cell by cell comparisons of observed
with expected values indicates that some cells contain too many students while others contain too few.
The pattern of these "residuals" of observed from
expected values describes the kinds of associations
that exist among the dimensions and, by inference,
the factors that are associated in the student
population.
     To account for these residuals, the analyst
constructs a series of models that posit different

Table B.2   Sources of Variation in Data in Table B.1

Source of Variation	df	Meaning
A	1	Univariate marginals
O	2	
X	2	
AO	2	Bivariate partial
AX	2	associations
OX	4	
AOX	4	Interaction

A = attentiveness to science
O = occupational aspirations
X = educational status

sources of variation. It is possible, for instance,
that all of the observed variation in Table B.1
could be due to the fact that there are more students who are inattentive to science than there are
students who are attentive. If this were the case,
a model that specified that the overall ratio of
attentives to inattentives exists in each of the nine
remaining subsamples would fit the observed data.
Model H(2) in Table B.3 states this assumption by
the inclusion of a single marginal (A) in the model.
The difference between the L.R. χ^2 for model H(2) and
the L.R. χ^2 for model H(1) indicates that some of the
variation in the observed values is due to variation
in attentiveness, but there is still considerable
discrepancy between the observed and expected values
taking only this variation into account.
     Models H(3) and H(4) incorporate the assump-
tions that in addition to variation in attentiveness, students also vary in their occupational
aspirations and their educational statuses. Since

Table B.3   A Hierarchy of Log-linear Models for
Data in Table B.1

Models		LRX^2	df	Probability
H1	Null	3892.97	17	.0001
H2	A	2194.63	16	.0001
H3	A,O	647.58	14	.0001
H4	A,O,X	604.45	12	.0001
H5	A,O,X,AO	550.65	10	.0001
H6	A,O,X,AO,AX	241.43	8	.0001
H7	A,O,X,AO,AX,OX	19.28	4	.001
H8	A,O,X,AO,AX,OX,AOX	0.0	0	----

A = attentiveness to science
O = occupational aspirations
X = educational status

the L.R. χ^2 is reduced appreciatively with the addition of each of these assumptions, one knows that
incorporating information about variation on each of
these dimensions reduces the discrepancies between
observed and expected values among the subsamples in
Table B.1.
Setting these three marginals of the table
equal to the proportional representation of the students in the observed sample allows the analyst to
test for the existence of associations among the
dimensions. If all the variation among cells in the
table is due to variation in the three marginals,
the expected values in the cells are easily calcu-
lable as the product of the marginal proportions
times N:

$$F_{ijk} = p_i \; p_j \; p_k \; N, \qquad (2)$$

where F_{ijk} is the expected value for subsample described by the ith category of variable I, the jth
category of variable J, and the kth category of
variable K. Table B.4 gives these expected values
for the marginal distributions in Table B.1.
Comparing the expected values in Table B.4 with
the observed values in Table B.1, one can see that
there is a better fit than one would obtain by assuming an equal distribution of students across all
three marginals (i.e., expected values of 207.67 for
all subsamples). Differences remain, however, between the observed and expected values, suggesting
that the three marginals alone do not adequately
account for all the variation in the table.
The sources of variation that have been left
out of model H(4) and the corresponding equation (2)
represent associations among the dimensions of Table B.1. The absence of fit between the observed
and expected values leaving these associations out
suggests that one or more associations exist in the
population from which these data were drawn. To
determine which associations these are, the analyst
continues the process of constructing models posit-
ing the existence of certain associations and the
absence of others. In the present example, inclusion
of association between each of the pairs of marginals produces an additional decrease in the L.R. χ^2
indicating that the only adequate fit to the observed data is obtained when all possible sources of
variation are taken into account, including the interaction term AOX. The analysis of Table B.1,
therefore, indicates that (a) there are bivariate

associations between attentiveness to science and
both occupational aspirations and educational status, and (b) the strength of these associations vary
across subpopulations of students with similar
aspirations and similar statuses.

MULTIPLICATIVE AND LOG-LINEAR MODELS

Equation (2) above implies a multiplicative structure for the models that predict expected values.
Goodman (1972a) demonstrates that a multiplicative
model of the form:

$$F_{ijk} = \eta\, \tau_i^A \tau_j^O \tau_k^X \tau_{ij}^{AO} \tau_{ik}^{AX} \tau_{jk}^{OX} \tau_{ijk}^{AOX} \tag{3}$$

can be derived from the odds-ratios among category
frequencies such that there are as many tau parameters in the multiplicative model as there are
sources of variation in the contingency table being
analyzed. Moreover, each tau parameter will have a
unique interpretation referring to the relative
contribution of each source to the predictive accuracy of the model. Using this model, then, it is

Table B.4. Expected Values for Model H4

Group	Occupational Aspirations	Attentiveness to Science	
		Nonattentive	Attentive
High School-Non-College Bound	Undecided	195.820	42.059
	Medium/Low	650.228	139.658
	High	159.070	34.166
High School-College-Bound	Undecided	176.414	37.891
	Medium/Low	585.791	125.818
	High	143.306	30.780
College	Undecided	228.697	49.120
	Medium/Low	759.398	163.106
	High	185.777	39.902
Total		3084.500	662.500

possible to partition variation among a set of expected values (F_{ijk}) to identify the existence and
form of each likely source of variation in the population from which the sample was taken.

To simplify the interpretation of the model
stated in equation (3), one can take the natural
logarithms of the terms in the model and state the
result as an additive "log-linear model" of the
expected cell frequencies:

$$G_{ijk} = \theta + \lambda_i^A + \lambda_j^O + \lambda_k^X + \lambda_{ij}^{AO} + \lambda_{ik}^{AX} + \lambda_{jk}^{OX} + \lambda_{ijk}^{AOX} \quad (4)$$

where G_{ijk} = ln F_{ijk} ; θ = lnη ; and λ = lnτ . The
advantage of the log-linear over the multiplicative
form of the model is that the log-linear model is in
the form of the familiar general linear model, and
the terms of the log-linear model are interpretable
as analogues to the main-effect and interaction
terms that make up the general linear model. Terms
with a single superscript (e.g., A) represent variation on a single specified marginal. Terms with two
superscripts (e.g., AB) represent partial associations. Terms with more than two superscripts represent first-order and higher "interactions" (in analysis of variance terminology). Interactions specify
variation in the strength of association of any two
of the included marginals across subpopulations defined by cross-classification of the remaining marginals in the term.

The additive form of the log-linear model allows one to conceptualize the contribution of each
source of variation as that which is made "in addition to" the contribution of all other terms contained in the model. This conceptualization suggests the manner in which hierarchies of log-linear
models are employed to build hypothetical models of
sample populations and to test specific hypotheses
about marginal distributions, associations, and interactions in these populations.

MODEL BUILDING AND HYPOTHESIS TESTING

The goal of model building is to determine the
smallest number of parameters necessary to adequately reproduce the distribution of observed data among
subsamples, thereby describing a hypothetical population from which the data might have been sampled.

The goal of hypothesis testing is to determine
whether a specific marginal distribution, association, or interaction derived from prespecified
theories, exists in a given sampled population.
Both these goals can be accomplished simultaneously
through construction of a hierarchy of models and
identification of a parsimonious model that provides
a likely interpretation of the observed data.
    The hierarchy of models in Table B.3 was used
above to illustrate how one proceeds in building an
adequate model and testing the parameters contained
within it. That process applied to the data in
Table B.1 led to the conclusion that the "saturated"
model H(8) was the only model that adequately fit
the observed data. This conclusion implies that
each of the terms in H(8) makes a statistically
significant ($p < .05$) contribution to the prediction
of the observed data in addition to the contributions made by each other term in the model. Had
this not been the case, one of the other models in
the hierarchy H(1) through H(7) would have produced
a L.R. χ^2 small enough (when compared with the chi-
square distribution with the appropriate degrees of
freedom) to judge the model as a likely description
of the population from which the data were sampled.
In this case, the more parsimonious model would have
been selected, and the terms left out of this model
would be assumed to make no reliable contribution as
sources of variation in the population.
    Having identified a model that adequately de-
scribes the observed data, the contribution of each
term in the model can be measured by comparing the
L.R. χ^2 of models including the term with the L.R.
χ^2 of models that leave only that term out. The
contribution of the interaction AOX, for instance,
can be measured by subtracting $\chi^2 H(8) - \chi^2 H(7) =$
(19.28 - 0 = 19.28). Comparing this difference to
the chi-square distribution with df H(7) - df H(8)
degrees of freedom provides a measurement of the
statistical reliability of concluding that the AOX
interaction exists in the population.
    The relative contribution of each term is mea-
sured by dividing this same difference by the total
L.R. χ^2 that needs to be accounted for by associations among the dimensions, as indicated by the L.R.
χ^2 for the "independence" model H(4). In the text,
this measure is referred to as the "Coefficient of
Multiple-Partial Determination" (CMPD). For the
data in Table B.1, the interaction AOX has a CMPD of
(19.28/604.45 = .03) indicating that the AOX inter-

action contributes approximately 3 percent to the
reduction in discrepancy between observed and expected values over and above the contribution of all
the bivariate associations in the model.

It is clear from this example that, given the
large sample size of the NPAS survey, a term may
make a statistically reliable contribution to the
prediction of the observed data without producing a
substantial contribution to the accuracy of the prediction. In the present study, terms have been retained in the predictive models only if they meet
both statistical and substantive criteria. The effect of using both criteria is that the reported
models include all of the major contributors to
variation among the specified subpopulations but may
exclude some main-effects and interactions that suggest slight differences among certain subpopulations
that do not appear to be of substantive importance.

LOGIT MODELS

Much of the present analysis has focused on predicting and explaining variation in a single
variable -- attentiveness to organized science. A
subset of log-linear models called "logit" models
is particularly appropriate for this purpose.

Logit models segregate variation in a single
dependent variable from variation due to other variables in a table and partition this variation among
the predictors. The starting point for a hierarchy
of logit models is a model that "fits" (i.e., takes
into account) all the associations among the predictor variables plus variation due to the marginal
distribution of the dependent variable (see model
H(1) in Table B.5). The interpretation of this "independence" model for the logit is that the L.R. χ^2
contains only the variation in the dependent variable that is available to be accounted for through
associations between the dependent variable and the
predictors.

Variation in the dependent variable is then
attributed to associations with the predictors.
Interactions are included in the logit model up to
and including the highest order interaction as long
as the higher order interactions contribute reliably
and substantially to the prediction of the sample
data.

The logit models for Table B.1 (treating attentiveness as dependent) are given in Table B.5. The
independence model H(1) produces a L.R. χ^2 of 342.27.
The predictor variables -- occupational aspirations
and educational status -- make quite different contributions in predicting variation in attentiveness.
The Coefficients of Multiple-Partial (CMPD's) indicate that educational status accounts for 79 percent of the variation in attentiveness independent
of occupational aspirations, while occupational
aspirations account for only 4 percent of the variation independent of educational status. The interaction AOX makes a reliable but not very substantial
contribution over and above the additive contributions of the two predictors.
The total predictable variation in attentiveness is measured by taking the difference in L.R. χ^2
between the independence model and a given logit and
dividing this difference by the L.R. χ^2 associated
with the independence model:

Table B.5   Logit Models for Data in Table B.1

Models		LRX^2	df	CMPD
H1	OX,A	342.27	8	----
H2	OX,AO,AX	19.28	4	.94
H3	H2 minus AX	288.48	6	----
	Difference due to AX	269.19	2	.79
H4	H2 minus AO	33.05	6	----
	Difference due to AO	13.76	2	.04
H5	OX,AX,AO,AOX	0.0	0	1.00
H6	H5 minus AOX	19.28	4	----
	Difference due to AOX	19.28	4	.06

A = attentiveness to science
O = occupational aspirations
X = educational status

$$[\chi^2_{H(1)} - \chi^2_{H(2)}]/\chi^2_{H(1)}$$

$$= (342.27 - 19.28)/342.27 = .94$$

The result (.94) indicates that the bivariate associations of the predictors with attentiveness jointly account for 94 percent of the variation in attentiveness to science, the remaining 6 percent being
associated with the interaction AOX. This result,
referred to as the "coefficient of multiple determination" by Goodman (1972b), is analogous to the coefficient of multiple determination (R^2) that would
be obtained by using subsample means, rather than
measures on individuals, in non-linear multiple
regression including all the same terms as the logit
model.

PATH ANALYSIS

When the assumptions one is willing to make about a
causal system are consistent with the information
one is using to make predictions about variation in
a dependent variable, it is possible to attribute
causal import to the sources of variation that contribute to the prediction of a series of dependent
variables (Darlington 1968). The assumptions one
needs to be able to make are:

1. The causal order among the variables is
    known such that each variable can be
    specified unambiguously as either a de-
    pendent or a predictor variable within
    each model.

2. The variables in the system represent a
    closed causal system such that no addi-
    tional variables could be introduced
    which would change the inferred causal
    relationships among the variables in
    the model.

3. The measured indexes used to identify
    associations are highly reliable and
    valid indicators of the constructs to
    which they refer such that the measured
    associations are reasonable estimates
    of the underlying causal parameters.

While it is nearly impossible to satisfy all three assumptions for any set of real data, cautious use of predictive models to make causal inferences can be of heuristic value. In the present study, causal inferences have been made when it seemed reasonable to assume a clear causal order among the variables and it appeared that no variables had been left out of the models that could drastically affect the conclusions. Multiple item indices were used extensively in an attempt to clarify the construct validity of the measures and to ensure their reliability.

Logit models were employed to identify the existence of causal relationships (Feinberg 1977). By constructing a series of logit models, each specifying a different dependent variable in the system, it is possible to infer the "effects" of a set of predictors on a set of dependent variables. Logit models are used for identifying the causal parameters within each stage of the causal process so all the possible causal relations among the predictors can be absorbed in an interaction term and set equal to the observed associations.

Lambdas (the parameters of the log-linear and logit model), standardized by their standard errors $(S.V. = \lambda/S_\lambda)$, were used as indicators of the existence and form of association in path diagrams of the causal systems. Under the null hypothesis of no differences, standardized lambdas are normally distributed with mean zero and variance 1.0 (Goodman 1971). Hence, it is reasonable to assume that cell frequencies producing standardized lambdas of ± 2.0 or greater are unlikely to be drawn from populations in which there are no differences among subpopulations.

Fig. B.1 illustrates the use of standardized lambdas as measures of partial associations among the three variables in Table B.1. The path analysis is based on the assumption that the development of occupational aspirations and the attainment of current educational status are both causally prior to, and potentially causes of, the development of attentiveness to organized science. Logit model H(2) was used as the predictive equation to illustrate the identification of causal parameters.

The standardized lambdas indicate relationships between educational status and occupational aspirations and causal effects of both variables on attentiveness. College students are significantly more likely than chance to be aspiring to a high prestige

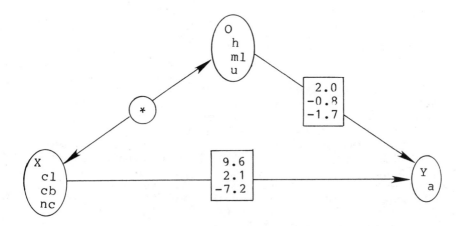

*	h	ml	u
cl	10.4	-1.1	-10.1
cb	6.9	-5.4	-3.3
nc	-12.6	-5.6	12.9

```
X  = educational status
  cl     college
  cb     high school-college bound
  nc     high school-non-college bound

O  = occupational aspirations
  h      high prestige
  ml     medium-to-low prestige
  u      undecided

Y  = attentiveness to organized science
  a      attentive
```

Fig. B.1. Path model of the relationships among
attentiveness to organized science, occupational
aspirations, and educational status.

who are undecided (S.V. = -1.7) or those who are aspiring to a medium prestige occupation (S.V. =
-0.8), while college students are more likely to be
attentive (S.V. = 9.6) than college bound high
school students (S.V. = 2.1), and the latter, in
turn, are more likely to be attentive to science
than are high school students who are not college
bound (S.V. = -7.2). Assuming the truth of the
underlying assumption regarding causal order, the
path diagram can be interpreted as indicating that
(a) both educational status and occupational aspirations contribute to the emergence of attentiveness
to science, with educational status accounting for
most of the variation, and (b) some of the impact
of each variable may be channeled through its influence on the other predictor.

Notes

CHAPTER 1

1. The term "organized science" as used in this
   analysis refers to both basic science and to
   technology. LaPorte (1978) has put forth a
   persuasive argument that science and technology
   are substantially different and that it is important to make a differentiation between them,
   especially for public policy purposes. The
   authors concur with LaPorte's point. At the
   same time, it is necessary to recognize that
   only a very small portion of the general population is able to understand or utilize the differentiation between science and technology, making
   it most difficult to construct a survey instrument that effectively assesses a differential
   attitude. In a separate survey of adults,
   Miller and Prewitt have proposed an instrument
   that utilizes case studies and other approaches
   to measure differential attitudes toward science
   and technology, but those approaches were not
   developed at the time the 1978 National Public
   Affairs Study was designed. Therefore we must
   utilize the broader concept for the purposes
   of this analysis.

CHAPTER 9

1. Eighty-five percent of the students reported an
    educational level completed for both father and
    mother. Fifty-eight percent of the parents in
    these families were reported to have exactly the
    same level of education within the four category
    classification -- less than high school, high
    school-vocational, bachelor's degree, and grad-
    uate-professional. Another 25 percent were re-
    ported in which fathers had completed more form-
    al education than mothers. The "educational
    environment" of most of the students was based
    on their father's education or reflected similar
    levels of education for both mother and father.
    For 154 students in the sample, "educational
    environment" is based on the student's report of
    only one parent's education. Given that stu-
    dents could have reported educational levels for
    parents who were absent from the family, the
    measure of "educational environment" undoubtedly
    underestimates the number of single-parent fam-
    ilies represented in the sample and, for these
    families, may overestimate the level of parents'
    education to which the student was actually ex-
    posed.

2. Fifty-two percent of the students reported an
    occupation for both parents. Another 1350
    students reported the "occupation" of their
    mothers as "homemaker." For the latter group
    and for the other 302 students who reported an
    occupation for only one parent, the occupational
    prestige of the home was classified on the basis
    of a single parent's occupation.

3. Family social status has been collapsed into two
    categories in Table 9.2, primarily on the basis
    of differences in parents' education. Families
    in which at least one parent completed a bache-
    lor's, graduate, or professional degree were
    classified as "higher status" unless neither
    parent held a medium or high-status occupation.
    The 216 families with college educated parents
    who hold low-status jobs were classified as
    "lower status" on the basis of the relatively
    low-prestige occupations (rated between 17 and
    40 on the Hodge, Siegel, Rossi scale). Families
    in which neither parent completed college were

classified as "lower status" unless at least one
parent held a high-status occupation (rated between 70 and 82 on the Hodge, Siegel, Rossi
scale). There were 201 families that were classified as "higher status" on the basis of occupational prestige.

4. The interaction between parents' education and
student's educational cohort is also significant
in predicting attentiveness (chi-square = 28.82,
df = 3, p<.001). The association between parents' education and attentiveness is different
for high school than for college students. Among
high school students, the proportion attentive
is highest (22 percent) among students from families in which the highest level of educational
attainment is a bachelor's degree. Among college
students, the proportion attentive is highest
(37 percent) among students from families in
which at least one parent holds a graduate or
professional degree. The general pattern, however, holds for each student cohort: students
from college educated families are more likely
to be attentive to science than are students from
families with less than a college education.

CHAPTER 10

1. It should be noted that the proportion of mutual dependence explained by each of the main
effects will not equal the combined explanatory
power of all of the main effects in combination
(H2 in the model reported in Table 10.1), since
some of the mutual dependence is jointly explained by more than one main effect and the
deletion procedure identifies only the unshared
explanatory power of each main effect. For a
more thorough discussion of this issue, see
Goodman (1972a, 1972b) or Fienberg (1977).

2. Students who reported that they were in the upper quarter of their class were considered high
on academic achievement and all other respondents were classified as low on the measure.

3. The science areas measure was dichotomized --
zero or one area versus two or more areas -- for
the logit model analysis.

CHAPTER 11

1. See discussion in chapter four for the ration-
    ale for using only the college sample in con-
    structing this and the other personality indices.

2. Twenty-three students have been dropped from the
    subsequent analysis because they could not be
    uniquely classified into one of the eight subsam-
    ples. Since "attentiveness to organized science
    and technology" is defined in this volume as
    combining interest, knowledge, and regular infor-
    mation acquisition in either science or techno-
    logy, certain patterns of responses were classi-
    fiable in terms of a single dimension or combin-
    ation of dimensions of attentiveness that were
    missing. Specifically, these response patterns
    were: (1) interest in science but not techno-
    logy, combined with knowledge of technology but
    not of science, and (2) interest in technology
    but not science, combined with knowledge of
    science but not of technology. In both pat-
    terns, the respondent indicated both interest
    and knowledge, but he or she would not be clas-
    sified as "attentive" to either science or tech-
    nology because interest and knowledge are not
    combined within either science or technological
    issues and concepts. As a result of dropping
    these 23 respondents, the proportion attentive
    to either science or technology in the following
    analysis is 0.1 percent greater than the propor-
    tion attentive in the total NPAS sample.

References

Abelson, P. H. 1975. "The Future of Chemical Information." In Communication of Scientific Information, ed. S. Day. Basel: Karger.

Adams, B. 1976. "Federal R and D Spending: The
Potential Impact of the Congressional Budget Pro-
cess." In Research and Development in the Federal
Budget: Colloquium Proceedings at American Asso-
ciation for the Advancement of Science, ed. W. A.
Blanpied and C. Lighthizer, Washington, D. C.

Adelson, J. 1971. The Political Imagination of the
Young Adolescent. Daedalus 100:1013-1050.

Adelson, J. and O'Neil, R. 1966. Growth of Political
Ideas in Adolescence. Journal of Personality and
Social Psychology 4:295-306.

Aiken, L. R. and Aiken, D. R. 1969. Recent Research
Attitudes Concerning Science. Science Education
53:295-305.

Allen, H., Jr. 1959. Attitudes of Certain High
School Seniors Toward Science and Scientific
Careers. Science Manpower Project Monographs.
New York: Teachers College, Columbia University.

Allport, G. 1954. The Nature of Prejudice. New
York: Anchor Doubleday.

Almond, G. A. 1950. The American People and Foreign
Policy. New York: Harcourt, Brace, and Co.

Astin, H. and Mvint, T. 1971. Career Development of
   Young Women During the Post-high School Years.
   Journal of Counseling Psychology 18:369-93.

Barber, B., Lally, J., Makarushka, and Sullivan, D.
   1973. Research on Human Subjects. New York:
   Sage Publications.

Baumel, H. B. and Berger, J. J. 1965. An Attempt to
   Measure Scientific Attitudes. Science Education
   49:267-69.

Beck, P. A. 1977. "The Role of Agents in Political
   Socialization." In Handbook of Political Socialization, ed. S. A. Renschon. New York: Free Press.

Belt, S. L. 1959. Measuring Attitudes of High School
   Pupils Toward Science and Scientists. Unpublished
   Ed.D. dissertation, Rutgers University.

Berelson, B. R., Lazarsfeld, P. F. and McPhee, W. N.
   1954. Voting: A Study of Opinion Formation During
   a Presidential Campaign. Chicago: University of
   Chicago Press.

Berger, P. L. and Luckmann, P. 1966. The Social
   Construction of Reality. Garden City, N. Y.:
   Anchor Books,Doubleday and Company, Inc.

Bernstein, B. 1971. Class, Codes, and Control, I:
   Theoretical Studies Towards a Sociology of Language. London: Routledge and Kegan Paul.

Bishop, Y. M. M., Feinberg, S. E., and Holland, P. W.
   1975. Discrete Multivariate Analysis: Theory and
   Practice. Cambridge, Mass.: The Massachusetts
   Institute of Technology Press.

Bixler, J. E. 1958. The Effect of Teacher Attitude
   on Elementary Children's Science Information and
   Science Attitude. Unpublished Ph.D. dissertation,
   Stanford University.

Boocock, S. S. 1972. An Introduction to the Sociology of Learning. Boston: Houghton-Mifflin.

Boynton, G. R., Patterson, S. C., and Hedlund, R.
   1969. The Missing Links in Legislative Politics:
   Attentive Constituents. Journal of Politics
   31:700-21.

Bradburn, N. M. and Sudman, S. 1974. Response Effects in Surveys. Chicago: Aldine.

Bridgeham, R. 1972. Ease of Grading and Enrollment
in Secondary School Science. I. A Model and Its
Possible Tests. II. A Test of the Model. Journal
of Research in Science Teaching 9:323-43.

Brooks. H. 1968. The Government of Science. Cam-
ridge,Mass.: The Massachesetts Institute of Technology Press.

Bush, V. 1945. Science: The Endless Frontier. Washington,D. C.: U. S. Government Printing Office.

Campbell, A., Converse, P. E., Miller, W. E., and
Stokes, D. E. 1960. The American Voter. New York:
John Wiley and Sons.

Carter, L. J. 1976. Nuclear Initiatives: Two Sides
Disagree on Meaning of Defeat. Science 194:811-12.

Chafee, S. H. with Jackson-Beeck, M., Durall, J., and
Wilson, D. 1977. "Mass Communication in Political
Socialization." In Handbook of Political Social-
ization, ed. S. A. Renschon. New York: Free Press.

Chafee, S. H., McLeod, J. M., and Atkin, C. K. 1971.
Parental Influences on Adolescent Media Use.
American Behavioral Scientist 14:323-40.

Chaffee, S. H., McLeod, J. M., and Wackman, D. B.
1966. Family Communication and Political Socialization Paper presented to the Association for
Educational Journalism, Iowa City, Iowa.

Cobb, R. W. and Elder, C. D. 1972. Participation in
American Politics: The Dynamics of Agenda Building. Boston: Allyn and Bacon.

Cohen, B. C.1959. The Influence of Non-Governmental
Groups in Foreign Policy-Making. Boston: World
Peace Foundation.

_____. 1973. The Public's Impact on Foreign
Policy. Boston: Little, Brown, and Co.

Coleman, J. S. 1961. The Adolescent Society: The
Social Life of the Teenager and Its Impact on
Education. New York: The Free Press of Glencoe.

Converse, P. 1964. "The Nature of Belief Systems in Mass Publics." In Ideology and Discontent, ed. D. E. Apter, pp. 206-61. New York: Free Press.

_____. 1970. "Attitudes and Non-Attitudes: Continuation of a Dialogue." In The Qualitative Analysis of Social Problems, ed. E. R. Tufte. Reading,Mass.: Addison-Wesley.

Coopersmith, S. 1967. The Antecedents of Self-Esteem. San Francisco: W. H. Freeman and Co.

Dahl, R. A. 1971. Polyarchy: Participation and Opposition. New Haven, Conn.: Yale University Press.

Darlington, R. B. Multiple Regression in Psychological Research and Practice. Psychological Bulletin, 69, 1968, 161-162.

Davis, I. C. 1935. The Measurement of Scientific Attitudes. Science Education 19:117-22.

Davis, J. A. 1961. Great Aspirations: The Graduate School Plans of America's College Seniors. Chicago: Aldine.

_____. 1965. Undergraduate Career Decisions: Correlates of Occupational Choice. Chicago: Aldine.

Davis, M. 1960. Community Attitudes Toward Fluoridation. Public Opinion Quarterly 23:474-82.

Davis, R. C.1958. "The Public Impact of Science." In The Mass Media. Ann Arbor, Mich.: Survey Research Center, Monograph No. 26, University of Michigan.

Dawson, R. E. 1973. Public Opinion and Contemporary Disarray. New York: Harper and Row.

Dawson, R. E., Prewitt, K. and Dawson, K. 1978. Political Socialization. 2nd ed. Boston: Little, Brown and Co.

Devine, D. 1970. The Attentive Public. Chicago: Rand McNally.

Doderlein, J. M. 1976. Nuclear Power, Public
Interest and the Professional. Nature
264:202-203.

Downs, A. 1957. An Economic Theory of Democracy.
New York: Harper.

Duncan, O. D., Featherman, D. L., and Duncan, B.
1972. Socioeconomic Background and Achievement.
New York: The Seminar Press.

Easton, D. 1953. The Political System: An Inquiry
into the State of Political Science. New York:
Knopf.

_____. 1965a. A Framework for Political Anal-
ysis. Englewood Cliffs, N. J.: Prentice-Hall.

_____. 1965b. A Systems Analysis of Political
Life. New York: Wiley.

Easton, D. and Hess, R. D. 1961. "Youth and the
Political System." In Culture and Social Character. ed. S. M. Lipset and L. Lowenthal, pp.226
-51. New York: Free Press.

_____. 1962. The Child's Political World.
Journal of Political Science 6:229-46.

Etzioni, A. and Nunn, C. 1974. The Public Appreciation of Science in Contemporary America. Daedalus
103,no.3:191-205.

Ewing, D. 1964. The Managerial Mind. Glencoe, Ill.:
Free Press.

Feinberg, S. 1977. The Analysis of Cross-Classified
Categorical Data. Cambridge, Mass.: Massachusetts
Institute of Technology Press.

Fenno, R. 1966. The Power of the Purse. Boston:
Little,Brown, and Co.

Field, J. O. and Anderson, R. 1969. Ideology in the
Public's Conceptualization of the 1964 Election.
Public Opinion Quarterly 33:389-98.

Flacks, R. 1970. "The Revolt of the Advantaged:
Explorations of the Roots of Student Protest."
In Learning About Politics: A Reader in Political

Socialization, ed. R. S. Sigel. New York: Random
House.

_____. 1967. The Liberated Generation: An
Exploration of the Roots of Student Protest.
Journal of Social Issues 23:52-63.

Fraser, J. 1970. The Mistrustful - Efficacious
Hypothesis and Political Participation. Journal
of Politics 32:444-49.

Gamson, W. A. 1966. Rancorous Conflict in Community
Politics. American Sociological Review 31:71-81.

_____. 1968. Power and Discontent. Homewood,
Ill.: Dorsey Press.

Garcia, F. C. 1973. Orientations of Mexican American and Anglo Children Toward the U. S. Political
Community. Social Science Quarterly 53:814-29.

Glaser, B. 1964. Organizational Scientists: Their
Professional Careers. Indianapolis: Bobbs-Merrill.

Glaser, W. 1960. Doctors and Politics. American
Journal of Sociology 66:230-45.

Glock, C. Y., and Stark, R. 1965. Religion and Society in Tension. Chicago: Rand McNally.

Goodman, L. A. The Analysis of Multidimensional
Contingency Tables: Stepwise Procedures and
Indirect Estimation Methods for Building Models
for Multiple Classifications. Technometrics,
13, 1971, 33-61.

_____. 1972a. A General Model for the Anal-
ysis of Surveys. American Journal of Sociology
77:1035-1086.

_____. 1972b. A Modified Multiple Regression
Approach to the Analysis of Dichotomous Variables.
American Sociological Review 37:28-46.

Gordon, C. 1971. Looking Ahead: Self-Conceptions,
Race and Family as Determinants of Adolescent Ori-
entation to Achievement. Washington, D. C.:
American Sociological Association.

Gosnell, H. and Merriam, C. E. 1924. Nonvoting.
   Chicago: University of Chicago Press.

Greenberg, D. 1977. Science and Government Report
   7:3, 8.

_____.1967. The Politics of Pure Science.
New York: Plume.

Greenberg, E. S. 1970. Political Socialization, ed.
   E. S. Greenberg, 1st ed. New York: Atherton
   Press.

Greenstein, F. I. 1960. The Benevolent Leader;
   Children's Images of Political Authority. American
   Political Science Review 54:934-43.

_____. 1965. Children and Politics. New Haven,
Conn.: Yale University Press.

Halvorson, H. O. 1977. Recombinant DNA Legislation--
   What Next? Science 198 (October 28).

Harman, H. H. 1967. Modern Factor Analysis.
   Chicago: University of Chicago Press.

Harris, L. 1973. The Anguish of Change. New York:
   Norton.

Hawkins, B. W., Marando, V. L. and Taylor, G.
   1971. Efficacy, Mistrust, and Political Participation: Findings from Additional Data and Indicators. Journal of Politics 33:1130-36.

Hennessy, B. 1972. "A Headnote on the Existence and
   Study of Political Attitudes." In Political Atti-
   tudes and Public Opinion, ed. D. D. Nimmo and
   C. M. Bonjean. New York: McKay.

Hero, A. O. 1959. Americans in World Affairs.
   Boston: World Peace Foundation.

_____. 1960. Voluntary Organizations in World
   Affairs. Boston: World Peace Foundation.

Hess, R. D. 1971. "The Acquisition of Feelings of
   Political Efficacy in Pre-Adults." In Social
   Psychology and Political Behavior: Problems and
   Prospects, ed. G. Abcarian and J. W. Soule.
   Columbus,Ohio: Merrill.

Hess, R. D. and Torney, J. V. 1967. The Development of Political Attitudes in Children. Chicago: Aldine Pub. Co.

Hilton, T. and Berglund, G. 1971. Sex Differences in Mathematical Achievement--A Longitudinal Study. Princeton, N. J.: Educational Testing Service.

Hirsch, H. 1971. Poverty and Politicization; Political Socialization in an American Sub-Culture. New York: Free Press.

Hirsch, W. 1968. Scientists in American Society. New York: Random House.

Hodge, R. W., Siegel, P. M., and Rossi, P. H. 1964. Occupational Prestige in the United States. American Journal of Sociology 70:286-302.

Hoff, A. G. 1936. A Test for Scientific Attitude. School Science and Mathematics 36:763-70.

Hoffer, E. 1958. The True Believer. New York: New America Library.

Holton, G. 1978. Epilogue to the Issue, 'Limits of Scientific Inquiry.' Daedalus 107,no.2:227-34.

Horniq, D. F. 1971. "The President's Science Adviser Talks to Physicists About Research." In Science In America, ed. J. C. Burnham. New York: Holt, Rinehart, and Winston.

Inkeles, A. 1955. Social Change and Social Character: The Role of Parental Mediation. Journal of Social Issues 11:12-23.

Jachim, A. G. 1975. Science Policy Making in the United States: The Batavia Accelerator. Carbondale: Southern Illinois University Press.

Jackman. M. R. and Jackman, R. W. 1973. An Interpretation of the Relation Between Objective and Subjective Social Class. American Sociological Review 38:569-82.

Jaros, D., Hirsch, H., and Fleron, F. 1968. The Malevolent Leader: Political Socialization in an American SubCulture. American Political Science Review 62:564-75.

Jennings, M. K. and Niemi, R. G. 1974. The Political Character of Adolescence: The Influence of Families and Schools. Princeton. N. J.: Princeton University Press.

Kadushin, C., Hover, J., and Tichy, M. 1971. How and Where to Find Intellectual Elite in the United States. Public Opinion Quarterly 35:1-18.

Kahn, P. 1962. An Experimental Study to Determine the Effect of a Selected Procedure for Teaching the Scientific Attitudes to Seventh and Eighth Grade Boys Through the Use of Current Events in Science. Science Education 46:115-27.

Katz, J. 1972. Experimentation with Human Beings. New York: Sage Publications.

Katz, E. and Lazarsfeld, P. A. 1955. Personal Influence: The Part Played by People in the Flow of Mass Communication. Glencoe, Ill.: Free Press.

Kelman, H. C. 1958. Compliance, Identification, and Internalization: Three Processes of Attitude Change. Journal of Conflict Resolution 2:51-60.

Key, V. O. 1966. The Responsible Electorate; Rationality in Presidential Voting, 1936-1960, with the assistance of Milton C. Cummings, Jr. Cambridge, Mass.: Belknap Press of Harvard University Press.

King. L. 1975. "The Future of Chemical Information." In Communication of Scientific Information, ed. S. Day. Basel: Karger.

Kirscht, J. P. and Knutson, A. A. 1961. Science and Fluoridation: An Attitude Study. Journal Of Social Issues 17:37-44.

Kohn, M. 1969. Class and Conformity. Homewood, Ill.: Dorsey.

Kornhauser, A. 1965. The Mental Health of the Industrial Worker. New York: John Wiley and Sons.

Ladd, E. and Lipset, S. M. 1972. The Politics of Academic Natural Engineers. Science 176:1091-99.

Lane, R. A. 1959. Fathers and Sons: Foundations of
Political Belief. American Sociological Review
24:502-11.

Langton, K. P. and Jennings, M. K. 1968. Political
Socialization and the High School Civics Curricu-
lum in the United States. American Political
Science Review 62:852-67.

LaPorte, T. 1978. "Indicators of the Public's Attitudes toward Science and Technology." Science
Indicators, 1972, 1974, 1976. A Review and Prospective Reflections. A paper presented at the
meeting of the Social Science Research Council's
Symposium on Science Indicators, May 1978.

LaPorte, T. and Metlay, D. 1975a. Public Attitudes
Toward Present and Future Technologies; Satis-
factions and Apprehensions. Social Studies of
Science 5:373-98.

_____. 1975b. They Watch and Wonder: Public
Attitudes Toward Advanced Technology. A report to
the National Aeronautics and Space Administration.

_____. 1975c. Technology Observed: Attitudes
of a Wary Public. Science 188:121-27.

Lowery, L. F. 1966. Development of an Attitude Measuring Instrument for Science Education. School
Science and Mathematics 66:494-502.

_____. 1967. An Experimental Investigation
into the Attitudes of Fifth Grade Students Toward
Science. School Science and Mathematics 67:569-79.

Luttberg, N. 1968. The Structure of Beliefs Among
Leaders and the Public. Public Opinion Quarterly
32:398-409.

Maccoby, E. 1968. "Moral Values and Behavior in
Children." In Socialization and Society, ed.
J. Clausen. Boston: Little, Brown.

Maccoby E. E. and Jacklin, C. N. 1974. The Psychology of Sex Differences. Stanford, Cal.:
Stanford University Press.

Marvick, D. and Nixon, C. 1961. "Recruitment Con-
rasts in Rival Campaign Groups." In Political

Decision-Makers, ed. D. Marvick. Glencoe, Ill.:
Free Press.

Marwich, D. 1968. "The Middlemen of Politics."
In Approaches to the Study of Party Organization.
Crotty, Ed. Boston: Allyn and Bacon. 314-374.

Mausner, B. and Mausner, J. 1955. A Study of the
Anti-Scientific Attitude. Scientific American
192:35-39.

McClosky, H. 1964. Consensus and Ideology in Amer-
can Politics. American Political Science Review
58:361-82.

McCready, W. C. and Greeley, A. M. 1976. The Ultimate Values of the American Population. Beverly
Hills, Cal.: Sage Publications.

McGuire, W. J. 1968. "Personality and Susceptibil-
ity Social Influence." In Handbook of Personality
Theory and Research, ed. E. F. Borgatta and W. W.
Lambert. Chicago: Rand McNally.

McLeod, J. M. and Chafee, S. R. 1972. "The Construction of Social Reality." In The Social Influence Processes, ed. J. T. Tedeschi, pp.50-59.
Chicago: Aldine.

Menard, M. W. 1971. Science: Growth and Change.
Cambridge,Mass.: Harvard University Press.

Milbrath, L. W. 1965. Political Participations:
How and Why People Get Involved in Politics.
Chicago: Rand MdNally.

Miller, D. R. and Swanson, G. E. 1958. The Changing
American Parent. New York: John Wiley and Sons.

Miller, J. D. 1978. "Selective Attentiveness; A
Conceptual Framework for Understanding Public
Attitudes Toward Organized Science." A paper presented to the Annual Meeting of the Society for
the Social Study of Science.

_____. 1977. The Attentive Public for Organ-
ized Science; A Retrospective Analysis of the
1958 SRC Science News Study. Report 77-1, Public
Opinion Laboratory, Northern Illinois University.

_____. 1976. "The Impact of Two Decades of Space Exploration on the Development of American Attitudes Toward Science and Technology." A paper presented to the NSF-Carnegie-Mellon University Conference on Retrospective Technology.

_____. 1975. The Development of Pre-Adult Attitudes Toward Environmental Conservation and Pollution. School Science and Mathematics, December, 729-38.

Mueller, J. E. 1973. War, Presidents and Public Opinion. New York: John Wiley and Sons.

Mulkay, M. 1976. The Mediating Role of the Scientific Elite. Social Studies of Science 6:445.

Myers, B. E. 1966. An Appraisal of Change in Attitudes Toward Science and Scientists and of Student Achievement in an Introductory College Chemistry Course Relative to the Students' Backgrounds in High-School Chemistry and Physics. Unpublished Ed. D. dissertation, Pennsylvania State University.

National Assessment of Educational Progress. 1971. National Assessment Report 4, 1969-1970 Assessment, Science: Group Results for Sex, Region, and Size of Community. Washington, D. C.: U. S. Government Printing Office.

_____. 1973. National Assessment Report 7, 1969-1970 Assessment, Science: Group and Balanced Group Results for Color, Parental Education, Size and Type of Community and Balanced Group Results for Region of the Country, Sex. Washington, D. C.: U. S. Government Printing Office.

National Opinion Research Center. General Social Surveys, 1972-1978: Cumulative Codebook, 1978.

National Science Board. 1973. Science Indicators-- 1972. Washington, D. C.: U. S. Government Printing Office.

_____. 1975. Science Indicators--1974. Washington, D. C.: U. S. Government Printing Office.

_____. 1977. Science Indicators--1976. Washington, D. C.: U. S. Government Printing Office.

Nie, N. and Anderson, K. 1974. Mass Belief Systems
   Revisited: Political Change and Attitude Structure. Journal of Politics 36:540-91.

Nelkin, D. 1974. "Threats and Promises: Negotiating
   the Control of Research." In The Role of Experts
   in a Nuclear Siting Controversy. Bulletin of the
   Atomic Scientists 30,no.9:29-36.

_____ . 1971. Nuclear Power and Its Critics.
   Ithaca: Cornell University Press.

_____ . 1977. Technical Decisions and Democracy.
   Beverly Hills, Cal.: Sage Publications.

_____ . 1978. Threats and Promises: Negotiating
   the Control of Research. Daedalus 107,no.2:191-210.

Noll, V. H. 1935. Measuring the Scientific Attitude.
   The Journal of Abnormal and Social Psychology
   30:145-54.

Oetzel, R. 1961. The Relationship Between Sex Role,
   Acceptance, and Cognitive Abilities. Unpublished
   M. Ed. Thesis.

Omerod, M. B. and Duckworth, D. 1975. Pupils Atti-
   tudes to Science: A Review of Research. Windsor,
   National Foundation for Educational Research, 1975.
   Research.

Oppenheim, K. 1966. Acceptance and Distrust: Attitudes of American Adults Toward Science. Unpublished Master's Thesis, University of Chicago.

Orfield, G. 1975. Congressional Power: Congress
   and Social Change. New York: Harcourt, Brace
   and Jovanovich.

Paige, J. M. 1971. Political Orientation and Riot
   Participation. American Sociological Review
   36:810-819.

Pierce J. R. and Tressler, A. G. 1964. The Research
   State: A History of Science in New Jersey.
   Princeton, N. J.: Van Nostrand.

Pool, I., Abelson, R. B., and Popkin, S. 1964. Candidates, Issues, and Strategies; A Computer Simu-

lation of the 1960 Presidential Election, Cambridge,Mass.: Massachusetts Institute of Technology Press.

Price, D. K. 1965. The Scientific Estate. Cambridge, Mass.: Belknap Press.

_____. 1972. Who Makes the Laws: Creativity and Power in Senate Committees. New York: Schenkman Publishing.

Prothro, J. and Grigg, C. 1960. Fundamental Principles of Democracy: Bases of Agreement and Disagreement. Journal of Politics 22:276-94.

Prothro, J. W. and Grigg, C. M. 1969. "Fundamental Principles of Democracy: Bases of Agreement and Disagreement." In Empirical Democratic Theory. ed. C. Cnudde and D. Neubauer. Chicago: Markham.

Ramo, S. 1970. Century of Mismatch. New York: McKay.

Reagan, M. D. 1969. Science and the Federal Patron. New York: Oxford University Press.

Remmers, H. H. 1957. High School Students Look at Science. Report of poll number 50 of the Purdue Opinion Panel. November 1957.

Rettig, R. A. 1977. Cancer Crusade: The Story of the National Cancer Act of 1971. Princeton N. J.: Princeton University Press.

Richardson, R. A. 1974. The Selling of the Atom. Bulletin of the Atomic Scientists, 30:8, 1974, 28-35.

Roback, T. 1974. "Occupational and Political Attitudes: The Case of Professional Groups." In Political Opinion and Political Attitudes. ed. A. Wilcox. New York: John Wiley and Sons.

Roback, H. 1977. "Congress and R and D Budgeting." In Research and Development in the Federal Budget: FY 1978. ed. W. D. Shapley, D. I. Phillips, and M. Roback. Washington D. C.: American Association for the Advancement of Science.

Roberts, D., Pingree, S. and Hawkins, R. 1974. Do the Mass Media Play a Role in Political Social-

ization? Australian and New Zealand Journal of
Sociology 11:37-43.

Robinson, J. P. and Shaver, P. R. 1973. Measures of
Social Psychological Attitudes. Ann Arbor, Mich.:
Institute for Social Research.

Robinson et al. 1973. Measures of Occupational Attitudes and Occupational Characteristics. Ann Arbor,
Mich.: Institute for Social Research.

Rokeach, M. 1960. The Open and Closed Mind. New
York: Basic Books.

Rokeach, M. and Eglash, A. 1956. A Scale for Measuring Intellectual Conviction. Journal of Social
Psychology 44:135-41.

Rosenau, J. 1974. Citizenship Between Elections.
New York: Free Press.

_____. 1963. National Leadership and Foreign
Policy: The Mobilization of Public Support. Princeton N. J.: Princeton University Press.

_____. 1961. Public Opinion and Foreign Policy.
New York: Random House.

Rosenberg, M. 1965. Society and the Adolescent Self-
Image. Princeton, N.J.: Princeton University Press.

Roshal, S. M., Frieze, I. and Wood, J. T. 1971. A
Multitrait-Multimethod Validation of Measure of
Student Attitudes Toward School, Toward Learning,
and Toward Technology in Sixth Grade Children.
Educational and Psychological Measurement
31:999-1006.

Rossi, A. S. 1965. "Barriers to the Career Choice
of Engineering, Medicine, or Science Among American Women." In Women and the Scientific Professions: The MIT Symposium on American Women in
Science and Engineering, ed. J. A. Matfield and
C. Van Aken. Cambridge, Mass.: Massachusetts
Institute of Technology Press.

Schwirian, P. M. 1968. On Measuring Attitudes Toward
Science. Science Education 52:172-79.

Sewell, W. H., Hauser, R. M., and Featherman, D. L. eds. 1976. Schooling and Achievement in American Society. New York: Academic Press.

Shen, B. 1975. "Scientific Literacy and the Public Understanding of Science." In Communication of Scientific Information. ed. S. Day. Basel: Karger.

Sigel, R. S. and Hoskin, M. B. 1977. "Perspectives on Adult Political Socialization -- Areas of Research." In Handbook of Political Socialization. ed. S. H. Renschon. New York: Free Press.

Sklair, L. 1973. Organized Knowledge. London: Hart-Davis, MacGibbon.

Sniderman, P. M. 1975. Personality and Democratic Politics. Berkeley, Cal.: University of California Press.

Sniderman, P. F. 1978. The Politics of Faith. British Journal of Political Science 8:21-44.

Stewart, I. 1948. Organizing Scientific Research for War. Boston: Atlantic, Little, Brown, and Co.

Struve, O. 1971. "An Astronomer Discusses His Science and the Need for Cooperative Research." In Science In America. ed. J. C. Burnham. New York: Holt, Rinehart, and Winston.

Suchner, R. W. 1977. The Structure of Dogmatic Thinking: A Short Scale for Measuring Three Dimensions of Closed-Mindedness. Unpublished manuscript, Northern Illinois University.

Sullivan, J. L., Piereson, J. E., and Marcus, G. E. 1978. Ideological Constraint in the Mass Public: A Methodological Critique and Some New Findings. American Journal of Political Science 22:233-49.

Taviss, I. 1972. A Survey of Popular Attitudes Toward Technology. Technology and Culture 13:606-21.

Taylor, J. 1973. The Scientific Community. London: Oxford University Press.

Tedin, K. L. 1974. The Influence of Parents on the Political Attitudes of Adolescents. American Political Science Review 68:1579-92.

Tolley, H. Jr. 1973. Children and War. New York:
Teachers College Press.

Toren, N. 1972. Social Work. Beverly Hills, Cal.:
Sage Publications.

Tullock, G. 1972. Toward a Mathematics of Politics.
Ann Arbor, Mich.: Ann Arbor Paperbacks, The
University of Michigan Press.

Verba, S., and Nie, N. H. 1972. Participation in
America: Political Democracy and Social Equality.
New York: Harper and Row.

Vitrogan, D. 1969. Characteristics of a Generalized
Attitude Toward Science. School Science and Mathematics 69:150-58.

Walster, E. E., Aronson, E. and Abrahams, D. 1966.
On Increasing the Persuasiveness of a Low Prestige
Communicator. Journal of Experimental Social
Psychology 2:325-42.

Weinberg, A. M. 1972. "Sociotechnical Institutes
and the Future of Team Research." in Scientific
Institutions of the Future. ed. P. C. Ritterbush.
Washington: Acropolis Books.

Weinburg, A. 1967. Reflections on Big Science.
Cambridge, Mass.: Massachusetts Institute of
Technology Press.

Weiner, N. 1948. Cybernetics or Control and Communication in the Animal and Machine. New York:
John Wiley and Sons.

Weiss, C. H. 1974. What America's Leaders Read.
Public Opinion Quarterly 38:1-22.

Weissberg, R. 1974. Political Learning, Political
Choice and Democratic Citizenship. Englewood
Cliffs, N. J.: Prentice-Hall.

Weissburg, H. 1976. Consensual Attitudes and Attitude Structure. Public Opinion Quarterly 40:349-59.

Whitley, R. 1976. Umbrella and Polytheistic Scientific Disciplines and Their Elites. Social
Studies of Science 6:471.

Withey, S. B. 1959. Public Opinion about Science
    and the Scientists. Public Opinion Quarterly
    23:382-88.

Wrightsman, L. S. 1970. "Parental Attitudes and Be-
    haviors as Determinants of Children's Responses to
    the Threat of War." In Learning About Politics:
    A Reader in Political Socialization, ed. R. Sigel.
    New York: Random House.

Ziegler, L. H. 1966. The Political World of the High
    School Teacher. Eugene, Oregon: Center for the
    Advanced Study of Educational Administration,
    University of Oregon.

Zuckerman, H. 1977. Scientific Elite. New York:
    Free Press.

Zurcher, L. A., Jr. and Monts, J. K. 1972. Political
    Efficacy and Political Trust, and Anti-Pornography
    Crusading: A Research Note. Sociology and Social
    Research 56:211-220.

Index

Academic achievement,
    201,203,208,225
  Index of, 204
  Related to attentive-
    ness, 204
Almond, G., v, 46-48
Amoeba, understanding of,
    93-4
Attentive public for
    foreign policy, 50-51
Attentive public for
    organized science,
  Expected attitudes, 51
  Information needs, 52
  Risk-benefit attitudes,
    51-52
Attentive publics, 46-47
  And electoral politics,
    295-99
  And referenda, 299-302
  Elitism issue, 293-95
  Size of, 290-92
Attentive efficacy, 223-
    25,243,254,257-59
  Index of, 224-25
  Related to attentive-
    ness, 224,249
Attentiveness to orga-
    nized science,
  Development of, 128-33
  Conceptual typology,
    119-22

Related to
  attentive efficacy,
    224,249
  educational aspira-
    tions, 172,198-99,
    258-59,278-86
  faith in people, 237-
    38,249,258-59
  family politicization,
    191-92,197-98,254-56,
    278-86
  occupational aspira-
    tions, 176-78,199
  occupational prefer-
    ence, 215,278-86
  religious attendance,
    185-86
  religious beliefs,
    186-189,198
  religious denomina-
    tion, 187-88
  self-esteem, 220-22,
    250,257-59
  sex-role socialization,
    181-82,197-98,278-86
  socioeconomic status,
    170,197-98
Structure of, 123-28
Attitude toward regula-
  tion of chemical pro-
  duction, 159-61

Attitude toward inter-
national scientific
competition, 158-61
Attitude toward space
exploration, 159-61
Attitudes, measurement
of, 68-69
Attitudes on energy
issues, 155-58
Belief systems, 21-22
Benefits of organized
science, 138-43
Bush, V., 7
Career interest index,
265
Citizenship between
elections, 45-46
Coefficient of multiple
determination, 331
Coefficient of multiple-
partial determination,
328
Community colleges, 57
Confidence in congres-
sional information,
109-10
Confidence in president-
ial information, 110
Degrees of freedom, 321-
22
Diffuse support, 34-35
DNA, understanding of,
93-4
Easton, D., 34
Educational aspirations,
170-74,194,202,216
Related to attentive-
ness, 172,198-99,258-
59
Energy, 7-8,10,13-14
Estrangement. See open-
and closed-mind.
Ethnocentrism. See open-
and closed-mind.
Evaluation apprehension,
55
Factor analysis, 66-69
Factor scales, 67
Faith in people, 235-38,
245

Index of, 236-37
Related to attentive-
ness, 237-38,249,258-
59
Family discussion. See
family politicization.
Family influences, 167-
68,202,250
Family interest, 265
Family politicization,
190-92,194,202
Index of, 191
Parental transmission
of issue attitudes,
42-43
Related to attentive-
ness, 191-92,197-98,
254-56,277
Related to political
salience, 267-69
Federal government, 4-12
Congressional appropri-
ation process, 12-13
Spending preferences,
161-64
Flouridation, 25
Gamma coefficient, 66-67
Gamma matrices, 67
General attitudes toward
organized science,
136-38
General political inter-
est, 264-66
Hypothesis testing, 328-
29
Information acquisition,
99,241-246,249
Confidence in sources,
108-13
Index of, 56
Measure of, 114-16
Interest in organized
science, 241-46
Development of, 84-8
Index of, 55
Measurement of, 74-83
Issue complexity, 44-45
Issue decision-makers,
49-50
Issue elites, 47-48

Issue interest, 73-4
Issue literacy, 21
Issue specialization, 33, 43
Knowledge of science, 93-8,241-246,249
Index of, 56
Lambda, 327
Standardized, 332-34
Life Goals, 212-13,216
Index of, 213
Related to attentive- ness,214
Likelihood-ratio chi- square, 323
Logit models, 329-31
Log-linear models, 321, 326-27
Manhattan Project, 7,12
Media consumption, 99-108
Mobilized publics, 48-49
Model building, 328-29
Molecule, understanding of, 93-4.
Multi-item indices, 65- 66
National Association of Science Writers(NASW), 24
National Institutes of Health (NIH), 13
National Public Affairs Study (NPAS), 55-64
Missing data, 64-65
Participation rate, 61- 62
Population, 57
Sample, 57-60
Weighting sample, 62-63
National Science Board (NSB), 26
News magazine readership, 103-4
Newspaper readership, 100-1
Nuclear power issue, 96-7
Occupational aspirations, 175-80,194
Index of, 175-76
Related to attentive-

ness, 176-78,199
Occupational preferences, 215,217,258
Index of, 215
Related to attentive- ness, 215,276
Office of Scientific Re- search and Development (OSRD), 7
Open- and closed-mind, 225-34,243-44
Indices of, 68-69,228- 30
Related to attentive- ness, 231-35,249-50, 254,255
Organic chemical, under- standing of, 93-4
Organized science, 3-6,9, 16-17
Definition of, 3
Growth rate, 10
Priority setting, 13-14
Regulation of, 12,15
Path analysis, 331-34
Peer discussion. See peer politicization.
Peer politicization, 201, 208-09
Index of, 210
Related to attentive- ness, 211,256-57,277
Related to political salience, 269-70
Personality traits, 219, 252
Related to attentive- ness, 239-40,246-61
Relations among, 238- 39,247
Related to political salience, 271-72
Preadult attitudes,
Development of, 38-39
Toward government, 35
Toward organized sci- ence, 28-29
Toward political parti- sanship, 40-41

Toward the political
   system, 36-38
Toward the presidency,
   35
Political attitudes,
   22-24
Political socialization,
   38-46
Public attitudes, 21-22
Toward foreign policy,
   30-31
Toward low-saliency
   topics, 29-31
Toward organized sci-
   ence, 24-25
Toward politics, 33-34
Public opinion, 21-22
Qualitative variables,
   64-67,321
Religious attendance,
   185-86
Religious beliefs, 183,
   194
Index of, 183-84
Related to attentive-
   ness, 186-89,198
Religious denomination,
   187-88
Risk-benefit assessment,
   136-38,149-54
Risks of organized sci-
   ence, 143-49
Rokeach, M., 68
School cohort, 172,194
School politicization,
   208-09,216
Index of, 210
Related to attentive-
   ness, 210
Related to political
   salience, 269-70
Science instruction, 201,
   208
Index of, 204
Related to attentive-
   ness, 206-07
Science literacy, 20
Scientific information
   explosion, 5
Scientific specializa-

tion, 6
Scientific thinking,
   19-20
Self-esteem, 220-22,243,
   254
Index of, 221-22
Related to attentive-
   ness, 220-22,250,257-
   59,277
Sex-role socialization,
   180-82,194,202,216,
   253-54
Related to attentive-
   ness, 181-82,197-98,
   274-75,278-84
Single-mindedness. See
   open- and closed-mind.
Socioeconomic status,
   168-70,194
Components of, 168
Index of, 169
Related to attentive-
   ness, 170,197-98
Space exploration, 7-9,13
Specific support, 34-35,
   39-40
Sputnik, 7,12
Television viewing, 106
Trust in people. See
   faith in people.

About the Authors

JON D. MILLER is Associate Dean of the Graduate
School and an Associate Professor of Political
Science at Northern Illinois University. He
holds a Ph.D. in political science from North-
western University. Dr. Miller has served as
Director of Research Services and Assistant
Professor of Political Science at Chicago State
University, Director of Research for the Amer-
ican Hospital Association, and Associate Exec-
utive Director of President Lyndon Johnson's
National Advisory Commission on Health Manpower.
He has published in political science, science
education, and health care journals.

ROBERT W. SUCHNER is an Associate Professor of
Sociology at Northern Illinois University. He
received his Ph.D. in sociology and educational
psychology from the University of Wisconsin -
Madison in 1972. Dr. Suchner held the position
of Honorary Research Associate to Professor
Herbert C. Kelman at Harvard University during
1979. He has published articles on sex-role and
personality influences on attitudes, interests,
and opinions in various social psychology
journals.

ALAN M. VOELKER is a Professor of Curriculum and
Instruction (Science Education) at Northern
Illinois University. He is currently chair-
person of the faculty of elementary education.

Dr. Voelker received his Ph.D. in Science Education with a minor in chemistry from the University of Wisconsin - Madison in 1967. He spent four years as a principal investigator at the Wisconsin Research and Development Center for Cognitive Learning, and he has been a principal investigator or co-investigator on several research and teacher education grants. Dr. Voelker has published articles and chapters in books dealing with attitudes toward science and the environment, science concept learning, and science teacher education.